abc's of ZENER DIODES

by

Rufus P. Turner, Ph.D.

HOWARD W. SAMS & CO., INC.
THE BOBBS-MERRILL CO., INC.
INDIANAPOLIS • KANSAS CITY • NEW YORK

FIRST EDITION

FIRST PRINTING—1974

International Standard Book Number: 0-672-21045-2
Library of Congress Catalog Card Number: 74-75282

Preface

This book explains in reasonably simple terms how zener diodes work and how to use them. A familiarity with these diodes often helps the technician and the design engineer to reduce circuit complexity; we hope that this book will help promote that familiarity.

The treatment is nonmathematical. While it has been drawn from several sources, it is also selective. Thus, the applications given in Chapter 4 have been selected on the basis of practicality from a much more numerous group. The text assumes the reader has a working knowledge of conventional semiconductor diodes, as well as an acquaintance with elementary semiconductor theory.

This book is offered with the sincere hope that it will serve a number of immediate needs for information, while also acting as an idea generator.

RUFUS P. TURNER

Contents

CHAPTER 4

CHAPTER 5

APPENDIX

CHAPTER 1

Zener Diode Fundamentals

The zener diode (also called *avalanche diode, breakdown diode, reference diode,* and *regulator diode*) is named for Dr. Carl Zener, who did landmark work in the field of electrical breakdown in solids. As early as 1934, Dr. Zener advanced his theory of "electrical breakdown in solid dielectrics," presenting his findings and conclusions in a paper of that title in the *Proceedings of the Royal Society* (London, England).

The present zener diode is a bonus from silicon-diode technology, and it provides, through its exploitation of nondestructive breakdown, a relatively simple means of ac and dc regulation, device protection, dynamic coupling and bypassing, the supply of a reference voltage, and a host of kindred applications. In some areas of use, the zener diode supplies at low voltages action which is comparable to that supplied by the gas tube at higher voltages.

A complete physical and mathematical analysis of zener diode action is too complex for presentation here. In its place, a simplified —but, we hope, not oversimplified—explanation of this diode and its operation is given in practical terms below. For a more advanced treatment, the reader is referred to the present extensive literature on semiconductor physics.

1.1 DIODE CONDUCTION CHARACTERISTICS

In a junction-type semiconductor diode, reverse current flows when the n layer of the diode is biased positive and the p layer negative. This reverse current consists of the flow of *minority* carriers across the junction and therefore is small, compared with the forward current of the diode. In a silicon junction diode, reverse current is very low (junction reverse resistance very high).

Fig. 1-1 shows the typical conduction characteristic of a semiconductor diode. For our subsequent discussion, the forward portion of the curve is unimportant and it is dashed accordingly in Fig. 1-1. The conduction curve shows clearly that reverse current (microamperes) is considerably lower than forward current (milliamperes), except after point BV.

1.2 REVERSE BREAKDOWN

From Fig. 1-1, we see that the reverse current is extremely low and that it remains almost constant at a few microamperes as the

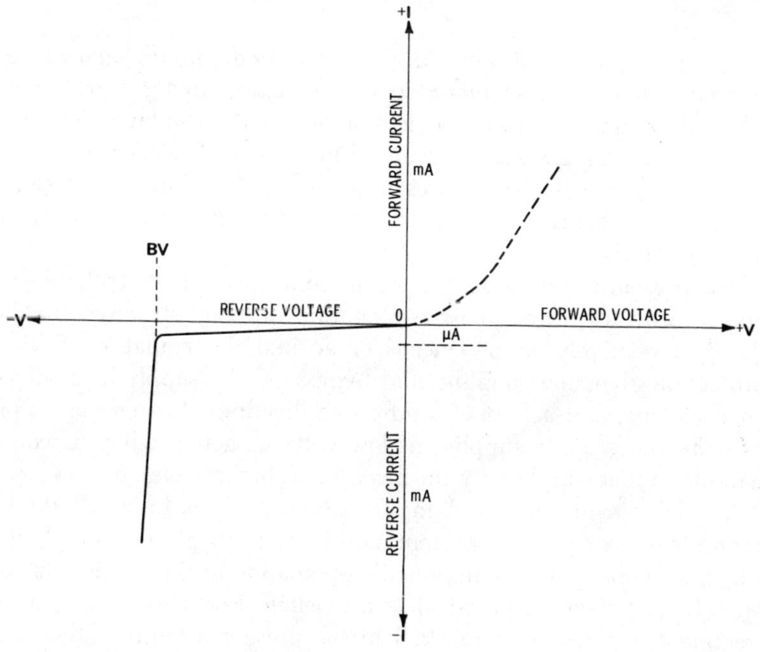

Fig. 1-1. Generalized diode conduction characteristic.

reverse voltage is increased from zero to a reasonably high negative value. However, after a particular critical value, BV, of reverse voltage is reached, the reverse current suddenly increases by a sizeable amount into the region of several milliamperes as the reverse voltage is further increased by a very small amount. For a small increment of reverse voltage beyond BV, the corresponding reverse-current increment is large. The converse also is true: For a large increment of reverse current, the corresponding increment of voltage drop across the diode is small.

Voltage BV is termed *breakdown voltage*. The point BV (see Fig. 1-1) at which the characteristic changes abruptly from low current (high resistance) to high current (low resistance) is also termed *breakover voltage* or *zener point*. It is easy to see why the word "breakdown" has been applied here, since the sudden steepening of the current slope resembles the comparable phenomenon of rapid current rise in a solid material, such as a dielectric, at the point of rupture, or breakdown.

Operation of the diode in its reverse-breakdown region—that is, along the steep slope of the curve, beyond BV—can destroy the junction unless the current is limited to a safe value. But when an operating point is chosen such that the rated power dissipation of the diode is not exceeded, a diode that is reverse-biased in this manner is useful for a number of special applications, and the breakdown is nondestructive. Ordinarily, however, conventional junction diodes are not operated in this manner.

A special diode which is manufactured expressly for operation in its reverse-breakdown region in such circuits as voltage regulators and voltage references is generically termed a *breakdown diode*. A variety of breakdown diodes may now be obtained commercially. Fig. 1-2 shows the circuit symbol (sometimes, the anode is labeled +

Fig. 1-2. Zener diode circuit symbol.

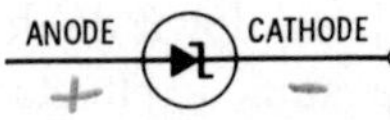

and the cathode −, but usually the symbol is used without letters or signs). To be very specific, we should mention that a breakdown diode in which BV is 5 V or lower is called a *zener diode*, and one in which BV is more than 5 V is called an *avalanche diode*. However, a custom has developed of calling all breakdown diodes *zener* diodes, and for simplicity this book follows that practice. In a typical "avalanche diode," one side of the junction is heavily doped,

whereas in a typical "zener diode," both sides of the junction are heavily doped.

1.3 ZENER BREAKDOWN AND AVALANCHE BREAKDOWN

Because of the narrowness of the junction, the electric-field intensity across a reverse-biased pn junction can be quite high, though the dc reverse voltage itself is relatively low. When breakdown voltage BV is under 5 V, breakdown results from the breaking up of covalent pairs in the doped-silicon atoms by this intense field and from the resulting ionization. This is *zener breakdown.* When, on the other hand, breakdown voltage BV is higher than 5 V, breakdown results from the breaking up of covalent bonds by minority carriers, which have been accelerated across the junction with sufficient momentum to collide forcibly with the crystal atoms. The collisions release additional minority carriers, which join the original ones and cause more collisions which, in turn, release still more carriers. Consequently, the current increases cumulatively. This is *avalanche breakdown.*

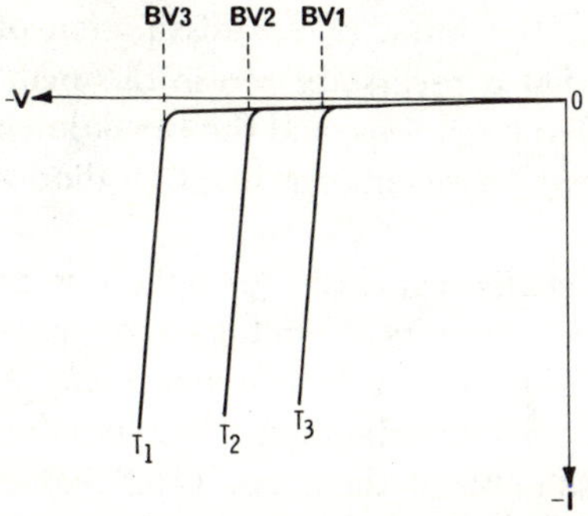

Fig. 1-3. Variation of breakdown voltage with temperature.

In some breakdown diodes, breakdown is due to a combination of zener and avalanche effects. And it may well be that both effects are present to some degree in all breakdown diodes.

1.4 TEMPERATURE DEPENDENCE OF BREAKDOWN

Breakdown voltage BV varies inversely as the temperature of the junction. That is, BV decreases as the temperature increases, and vice versa, this action describing a negative temperature coefficient of voltage. Fig. 1-3 illustrates this action. Observe that as the junc-

tion temperature increases from T_1 to T_2 to T_3, the breakdown voltage decreases from BV_3 to BV_2 to BV_1.

This temperature effect may be minimized in some applications by operating the zener diode on a suitable heat sink or, in some instances, by including compensating components in the circuit, as explained later under "Temperature Coefficient" in Section 1.7.

Internal temperature, as well as external ambient temperature, can adversely affect operation of a zener diode. Excessive power dissipation by the junction, for example, raises the temperature of the diode. Thus, the case temperature of the device is a function of both ambient and junction temperatures.

1.5 POWER DISSIPATION

The average power dissipated by the zener junction is equal to IV watts, where I is the average reverse current (amperes) and V the average reverse voltage (volts). In the breakdown region where current is high, this power can become excessive unless the current is properly limited to a safe value; excessive power dissipation can overheat or destroy the junction, as well as degrade circuit performance.

Manufacturers of zener diodes specify the maximum safe operating power, and derating curves may be employed to determine the extent to which power must be reduced at higher-than-recommended temperatures. Such curves usually are linear.

Fig. 1-4 shows the reverse zener characteristic with dashed curves (power hyperbolas) superimposed to show power dissipation at two points, X and Y. A power hyperbola indicates a constant value of power dissipation for a number of consecutive EI products. Thus, at any point along curve P2 the power may be, say, ½ watt, though the current values intersecting P2 will be different. If hyperbola P1 indicates maximum permissible power dissipation, and P2 indicates

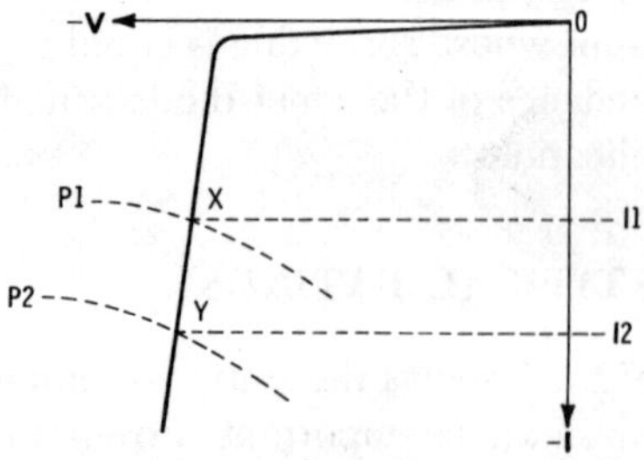

Fig. 1-4. Zener curve with power hyperbolas.

excessive dissipation, then zener current for the subject diode must be limited to I1 or lower at the specified voltage.

1.6 DC RESISTANCE AND AC IMPEDANCE

The dc resistance at any point along the reverse volt-ampere curve of the zener diode is determined from Ohm's law:

$$R_r = \frac{V_r}{I_r}$$

where,

R_r is the reverse resistance (ohms),
V_r is the reverse voltage (dc volts),
I_r is the reverse current (dc amperes).

It is clear from this relationship and the conduction curve that the resistance spans a wide range of values as conduction changes abruptly from low current before BV to high current after BV.

An example will serve to illustrate this action. A certain 1N3686 zener diode is operating in its breakdown region, at 20 V. At a lower value, 15.2 V, the reverse current is 5 μA, these values corresponding to a junction resistance of 3 megohms. At 20 V, however, the current is 6.2 mA, these latter values corresponding to a junction resistance of 3200 ohms. The resistance change here is about 940 to 1. This diode would be operating within its power rating (¾ watt), since here $P_D = 20(0.0062) = 0.12$ W.

The ac impedance of a zener diode operating in its breakdown region is considerably lower than the dc resistance at the same reverse-current point. For the 1N3686 unit described above, for example, the impedance is of the order of 25 ohms at 6.2 mA zener-region operation. Thus, the ac impedance at the 6.2-mA dc operating point is 0.78% of the dc resistance at that point. The ac impedance is determined by superimposing upon the direct current a 60-Hz current whose rms value is equal to 10% of the dc value. The low ac impedance of the zener diode suits this device to numerous dynamic applications.

1.7 TYPICAL RATINGS

Fig. 1-5 shows the static volt-ampere conduction curve for a zener diode. Various important current and voltage points are marked, and

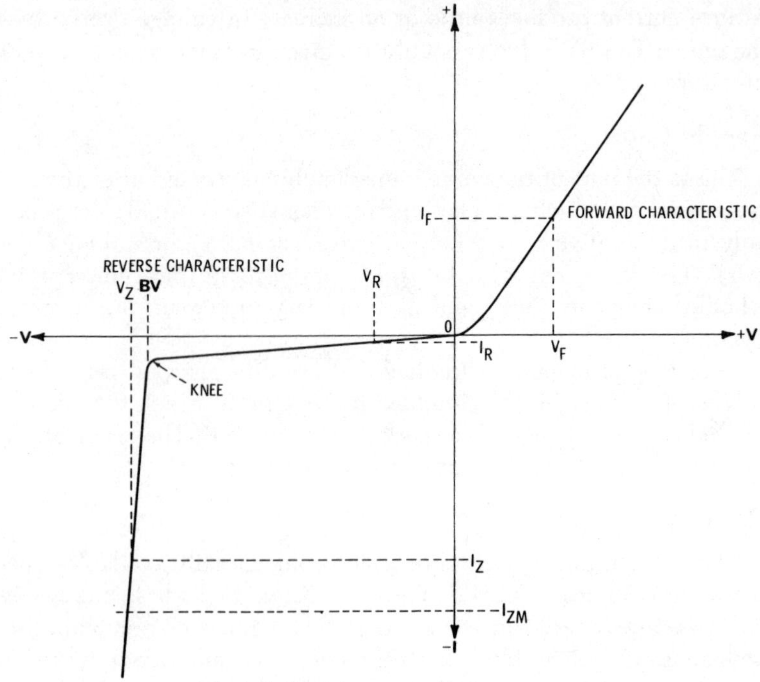

Fig. 1-5. Static zener conduction curve.

these ratings are discussed separately below. Unless otherwise noted, ratings are given for diode operation at 25°C.

Forward Characteristic

The forward conduction characteristic appears in the first quadrant in Fig. 1-5 and is similar to the forward characteristic of any semiconductor diode. However, since zener diodes are customarily operated only in the reverse region, the forward characteristic usually is of no interest, and most diode manufacturers do not specify forward-operating values. When a forward value is given, it is usually the forward voltage drop (V_F volts) across the diode for a selected test value of forward current (I_F milliamperes), as shown in Fig. 1-5.

Breakdown Voltage (BV)

This is the distinct value of reverse voltage at which the reverse current suddenly starts a rapid increase versus voltage; or, from the opposite point of view, it is the point at which a large increase in

reverse current produces little or no increase in voltage drop across the diode. The BV value is not always given in diode manufacturer's literature.

Knee of Curve

This is the part of the curve immediately before and after the discrete breakdown-voltage (BV) point. This knee is usually described only in qualitative terms, i.e., as *sharp* (or *hard*) or *rounded* (or *soft*). The knee may also be *stable* or *unstable*, depending upon whether or not its shape and position vary or remain fixed during operation.

Deserving of mention is the fact that even the sharpest zener knee is seen to be recognizably rounded if the scale to which the curve is plotted is reduced enough to show small variations. The zener breakdown is not so abrupt as that of a gas tube.

Nominal Zener Voltage (V_Z)

This is the rating by means of which commercially available zener diodes usually are identified. Thus, an "8.2-volt" diode is one whose zener voltage is specified as 8.2 V dc. The tolerance on nominal zener voltage is ±2%, 5%, 10%, or 20%, depending upon diode type and manufacture. Thus, a 5% diode advertised as 9.1 V nominal may have an actual zener voltage anywhere within the limits of 8.64 V and 9.55 V. Similarly, the same 9.1-V unit with 20% tolerance will have an actual V_Z value between 7.28 and 10.92 V.

A glance at Fig. 1-5 shows the meaning of V_Z. This is the voltage drop across the diode at a specified value of test current (I_Z in mA) in the zener breakdown region. This region is the one in which the diode is operated in all of its unique applications. Zener voltage V_Z should not be confused with breakdown voltage BV discussed earlier in this section and also shown in Fig. 1-5.

Zener diodes are commercially available with nominal V_Z ratings in steps between 0.77 V (e.g., Type 1N3896) and 580 V (e.g., Type 1N580). Typical V_Z values are 1.8, 3.6, 5.1, 5.6, 6.2, 9.1, 10.5, 12.8, 15.8, 18, 20, 27, 33, 39, 47, and 58 volts. (These values are often used in commercial equipment.)

Zener Current (I_Z)

This is the direct-current level (in mA) at which the diode is operated in the zener breakdown region. In the diode manufacturers' literature, it is the test current (usually labeled I_{ZT}) for which the

nominal zener voltage, V_Z, is specified; but it is not necessarily the highest current at which the diode may be operated without exceeding maximum power dissipation (see I_Z and V_Z in Fig. 1-5).

Maximum zener current (I_{ZM} in Fig. 1-5) is the highest value of direct current (in milliamperes) the diode will pass without exceeding the maximum permissible power dissipation. Test-current I_{ZT} values specified in the diode manufacturers' literature cover the range 0.1 mA (e.g., Type 1N1946) to 3200 mA (e.g., Type 1N4557B).

Reverse Leakage Current (I_R)

This is the small, reverse, direct current that flows when the diode reverse voltage is small enough to be safely below the zener knee (see I_R at reverse test voltage V_R in Fig. 1-5).

Typical values of reverse leakage current lie in the range 0.01 μA to 100 μA. Representative examples are 0.01 μA at 76 V (Type 1N4135) and 100 μA at 1 V (Type 1N4370). In Type 1N4135, the reverse leakage current of 0.01 μA at 76 V represents a reverse resistance of 7600 megohms.

Reverse leakage current, I_R, can increase rapidly with temperature. An I_R change of 100 to 1, for example, can occur between 25°C and 100°C at a given reverse test voltage, V_R.

Power Dissipation (P_D)

The dc power (watts) dissipated by the zener diode is equal to the product of zener voltage (V_Z volts) and zener current (I_Z amperes): $P_D = V_Z I_Z$. Excessive power dissipation can raise the junction temperature adversely and shift the operating characteristics of the diode. In extreme cases, the diode is destroyed.

Maximum permissible power dissipation (P_{DM}) values at 25°C cover the range 150 milliwatts (e.g., Type 1N228) to 50 watts (e.g., Type 1N2804A). Some diode manufacturers specify P_D at other temperatures as well, such as 50°C and 75°C. Power derating must be applied above a specified critical temperature, according to the curve or formula recommended by the manufacturer.

Operating Temperature Range

The range of junction temperature and storage temperature, given by zener diode manufacturers, extends from −65°C to +175°C.

The case temperature of the device is the total of junction temperature and ambient temperature. Above a certain critical temperature specified by the manufacturer, the diode power must be

derated (see under Power Dissipation, above). It should be mentioned, however, that some zener diodes are housed in metal cases—such as the DO-4, DO-5, and TO-3—that may readily be attached to heat-dissipating surfaces or devices and to which heat sinks may easily be fastened.

Temperature Coefficient (θ)

The temperature dependence of zener breakdown was discussed in an earlier section. Temperature compensation may be achieved in any of the conventional ways. In special zener diodes, built-in temperature compensation is secured by combining in the same unit a forward-biased junction in series with the negative-biased zener junction. Since the temperature coefficient of the forward junction has a sign opposite to that of the reverse junction, the two tend to cancel each other, and the net drift is low.

Specified temperature coefficients of zener current or zener voltage (in %/°C) cover the range 0.0001% (e.g., Type 1N4583) to 0.8% (e.g., Type 1N1742). Diode manufacturers state these temperature coefficients for a specified operating temperature range. Common temperature ranges (in °C) are 0 to 50, 0 to 70, 0 to 75, 25 to 100, −55 to +100, −55 to +125, and −55 to +150.

Zener Impedance (Z_Z)

As shown earlier in Section 1.6, the ac impedance of the zener diode is significantly lower than the dc resistance at a given dc zener current level. The small-signal impedance of the diode is measured by superimposing a 60-Hz signal on the zener current, as explained in Section 1.6, and measuring the resulting ac current. Zener impedance values cover the range 0.12 ohm (e.g., Type 1N4561) to 2500 ohms (e.g., Type 1N992). Zener impedance decreases as dc zener current is increased, and vice versa.

Junction Capacitance (C_J)

In common with other semiconductor devices, the reverse-biased zener diode exhibits a voltage-dependent junction capacitance. This characteristic is exploited in the varactor (voltage-controlled-capacitance diode), but is extraneous in the zener diode. Junction capacitance is highest at low dc reverse voltage. At a reverse voltage of 1 V, capacitance C_J is shown by diode manufacturers to be as high as 700 pF for a 3.9-V diode, and as low as 60 pF for a 100-V diode. But higher values can be found in occasional units.

Noise Density ***(N_D)***

White noise is generated by the reverse-biased zener diode. This noise amplitude (in rms microvolts per square root cycle) is specified by the diode manufacturer for a given dc test current, e.g., 250 μA. Typical values of N_D lie in the range 1 μV to 40 μV.

1.8 BASIC ZENER CIRCUIT

Fig. 1-6 shows the basic zener circuit. In practical applications of zener diodes, there are many variations of this simple arrangement.

Here, the diode (D1) is connected in series with a dc input source and limiting resistor R1. The polarity of the input voltage (E_i) is such that the cathode of the diode is positive; this causes reverse current (I_Z) to flow through the diode. The output voltage (E_o) of the circuit appears across the diode, being the voltage drop due to I_Z.

When resistance R1 is correctly chosen, the diode current is limited to a safe value within the steep breakdown region of the reverse-current characteristic. When this condition is fulfilled, a given change (ΔE_i) in input voltage produces a relatively large change (ΔI_Z) in zener current. But owing to the steepness of the slope of the reverse conduction curve in the breakdown region, a very small corresponding change (ΔE_o) occurs in the output voltage.

The opposite action also is obtainable. That is, if a fluctuating load is connected across the output terminals of the circuit, the zener current will decrease as the load current increases—and vice versa—and this action will maintain a constant total current through R1 and a constant value of load voltage E_o. Thus, the simple, basic zener circuit provides simple regulation against fluctuations in input

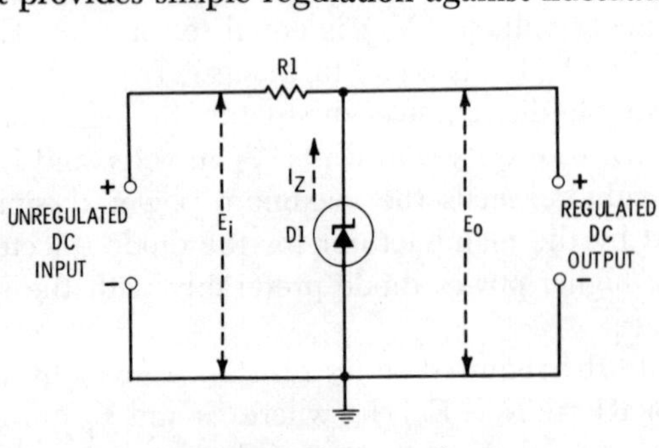

Fig. 1-6. Basic zener circuit.

voltage and/or output load. Applications other than voltage regulation depend upon this automatic limiting action provided by the zener diode.

Output voltage E_o, being the diode voltage drop, will correspond in most cases to the rated nominal zener voltage of the diode when current I_Z is limited by R1 to the value specified for V_Z by the diode manufacturer. Voltage-reference circuits (also identical with Fig. 1-6) depend upon this relationship.

1.9 DESIGNING THE ZENER CIRCUIT

Most applications of zener diodes utilize the regulating property (limiting action) of the basic zener circuit shown in Fig. 1-6. This circuit is employed principally as (1) a voltage regulator or (2) a voltage reference. In either case, the design procedure is essentially the same. The main advantages of the circuit are simplicity and economy; the arrangement consists of one diode and one resistor. In most instances, the arrangement is also space saving.

Voltage Regulator

The following step-by-step procedure is suggested for the design of a simple voltage regulator:

1. Determine the value of regulated output voltage (E_o) that is required.
2. Determine the value of unregulated input voltage (E_i) that is available.
3. Determine the maximum value of current (I_L) that will be drawn by the load.
4. From the zener diode manufacturer's literature, select a diode whose zener voltage (V_Z) is equal to, or near, E_o and whose zener current (I_Z) is equal to, or near, I_L.
5. Calculate the diode power dissipation by means of the formula $P_D = V_Z I_Z$, where P_D is in watts, V_Z in volts, and I_Z in amperes. If this value exceeds the maximum power dissipation recommended by the manufacturer for the diode selected in Step 4, select a higher-power diode preferably with the same V_Z and I_Z ratings.
6. Calculate the required series resistance (R1) by means of the formula $R1 = (E_i - E_o)/I_Z$, where E_i and E_o are in volts, I_Z is in amperes, and R1 is in ohms. (To determine the power dis-

sipation of the required series resistance, use the formula $P = I^2R$.)

Note that for any case, the dc supply voltage (E_i) must be somewhat higher than the desired regulated output voltage (E_o). Specifically, $E_i = E_o + I_Z R1$. Under conditions of fluctuating load, the voltage drop across the resistor is constant, because the diode current fluctuates inversely, thereby maintaining a constant current ($I_Z + I_L$) through this resistor. Therefore, the load voltage (output voltage E_o) remains constant within practical limits imposed by the individual diode characteristics. Voltage regulation of the simple circuit is good and is calculated from the formula:

$$\text{Percent regulation} = 100\frac{E_{oNL} - E_{oFL}}{E_{oFL}}$$

where,

E_{oNL} is the no-load value of output voltage E_o,
E_{oFL} is the full-load value of output voltage E_o.

ILLUSTRATIVE EXAMPLE: A certain transistor circuit requires a regulated dc voltage of 12 V at 50 mA. An unregulated supply of 18 V is available. Select a usable zener diode and calculate the power dissipation, required series resistance, and power dissipated by the resistor.

A search of diode manufacturers' literature discloses Type 1N1594, which is rated at $V_Z = 12$ V and $I_Z = 50$ mA. These current and voltage values, which correspond to the load requirements given above, give a power dissipation value $P_D = 12(0.05) = 0.6$ watt (see Step 5). Since the manufacturer rates the diode at 3.5 W, this operating power level is safely within limits.

The required series resistance, calculated according to Step 6, is $R1 = (18 - 12)/0.05 = 6/0.05 = 120$ ohms. Power dissipated by this resistor is $I^2R = 0.05^2(120) = 0.3$ watt.

Voltage Reference

A voltage-reference unit functions much the same as the simple regulator unit and has the same configuration (Fig. 1-6). However, its use differs principally in the fact that very little or no current is drawn from a typical reference unit, so that loading and load fluctuations are usually of no interest. In many applications, typified by potentiometric measurements and instrument calibration, the zener-

reference unit replaces the standard cell usually employed in these "no-load" measurements. Temperature-compensated zener *reference diodes* are commercially available, and these afford higher stability than is obtainable with conventional zener diodes. With reference to Fig. 1-6, the following step-by-step procedure is suggested for the design of a simple voltage reference:

1. Determine the value of reference voltage (E_o) required.
2. Determine the value of supply voltage (E_i) which is available.
3. From the zener diode manufacturer's literature, select a diode whose zener voltage (V_Z) is equal to, or near, E_o.
4. Note the corresponding value of zener current (I_Z) given in the literature.
5. Calculate the required series resistance by means of the formula $R_1 = (E_i - V_Z)/I_Z$, where E_i and V_Z are in volts, I_Z is in amperes, and R_1 is in ohms.

As in the simple voltage regulator, the dc supply voltage (E_i) in this circuit must be somewhat higher than the desired output reference voltage (E_o). In many applications, the output voltage of the reference circuit must be regulated against supply-voltage variations, rather than load fluctuations, and this it does excellently. Thus, when a battery is employed as the E_i source, the output voltage of the circuit remains constant during a long interval of battery deterioration.

ILLUSTRATIVE EXAMPLE: A 6.5-V dc reference-voltage source is required. The dc supply will be a 12-V battery composed of eight 1.5-V, size-C flashlight cells.

A search of the diode manufacturers' literature discloses Type 1N821A, which has a V_Z rating of 6.5 V at 7.5 mA and a temperature coefficient of 0.015%/°C (−55° to +100°C). This is a zener reference diode.

The required series resistance, calculated according to Step 5, is $R1 = (12 - 6.5)/0.0075 = 5.5/0.0075 = 733$ ohms. The power dissipated by this resistor is $I^2R = 0.0075^2(733) = 0.0000562$ $(733) = 41.2$ milliwatts.

Voltage Regulation Percentage

The voltage regulation of a zener regulator circuit may be calculated, as explained earlier, by means of the formula:

$$VR(\%) = 100\frac{E_{oNL} - E_{oFL}}{E_{oFL}}$$

where,

E_{oNL} is the no-load value of output voltage E_o,
E_{oFL} is the full-load value of output voltage E_o.

Illustrative Example: A voltage-regulator circuit employing a Type 1N3797 zener diode delivers a no-load output voltage of 22.5 V and a full-load output voltage of 22 V at 20 mA. Calculate the voltage regulation in percent.

$$VR(\%) = 100\frac{22.5 - 22}{22} = 2.27\%$$

CHAPTER 2

DC Power Supply Applications

Zener diodes are found in greater number and variety in dc supply than in any other field of application. This is a natural consequence of the unique zener characteristic, which is a dc phenomenon. The principal application is that of voltage regulation.

The zener diode itself can function as a regulator, or it can be used as a stable reference unit in a transistor- or tube-type regulator. Both types are described in this chapter. In the first case, an uncomplicated regulator is provided, which (mainly at voltages under 75 V) has the same simplicity as that of the gaseous regulator tube at higher voltages.

A comprehensive selection of basic circuits and practical circuits is presented below, each representing a range of possible applications of zener diodes. However, these circuits do not by any means exhaust the possibilities afforded by the many dozen types of zener diodes now commercially available.

2.1 SIMPLE ZENER-DIODE VOLTAGE REGULATOR

The simplest zener-diode dc voltage regulator consists of a diode connected in series with a current-limiting resistor. The unregulated voltage (E_i) is applied to the series combination, so poled as to bias the cathode of the diode positive, and the regulated output voltage

(E_o) is taken from across the diode. Fig. 2-1A shows this arrangement for grounded-negative output. Reversing the diode and the polarity of the input voltage provides grounded-positive output, as shown in Fig. 2-1B.

The simple circuit may be designed for individual operating requirements (available input voltage, desired regulated output voltage and current, and required series resistor R1) by means of the method explained in Section 1.9, Chapter 1. Table 2-1 lists 25 diode-

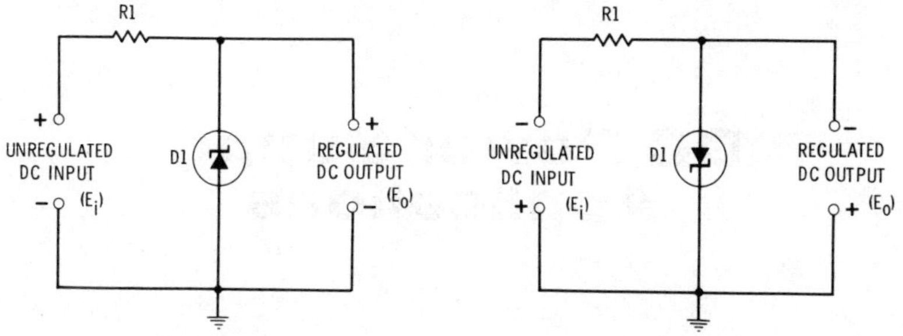

(*A*) *Grounded negative.* (*B*) *Grounded positive.*

Fig. 2-1. Simple zener diode voltage regulator.

resistor combinations for various output voltages from 1.8 to 200 V. These diodes and resistors were chosen for the practical output voltage and output current values shown; but these few presentations do not by any means exhaust the possibilities, there being a wide variety of commercially available diodes and numerous obtainable current and voltage combinations.

2.2 SERIES-DIODE CIRCUIT

When a higher regulated output voltage is desired than can be supplied by any single zener diode in the builder's possession, several diodes may be series-connected in place of the single diode shown in Figs. 2-1A and 2-1B. The sum of the individual V_Z values of the diodes must equal the desired output-voltage value. Thus, two 18-V diodes (such as Type 1N228) connected in series will give a regulated output of 36 V. Similarly, a 6.6-V, 10-mA diode (e.g., Type 1N2034) in series with a 4.4-V, 10-mA diode (e.g., Type 1N2032) will give an 11-V regulated output. The series-connected diodes should have identical zener-current (I_Z) ratings, although their zener-voltage (V_Z) values differ.

Table 2-1. Typical Diode/Resistor Combinations

Output Voltage (E_o, V)	Output Current (I_o, mA)	Input Voltage (E_i, V)	Series Resistance (R_1, Ω)	Type
1.8	20	3	60	1N370
3.5	10	6	250	1N373
6.2	100	9	28	1N1766
7.1	13	9	146	1N377
7.5	100	18	105	1N1768A
8	5	12	800	1N475
9.1	10	12	290	1N3517
10	50	18	160	1N1771A
10.5	10	18	750	1N701
10.5	500	22	23	1N2045
12	105	22	95	1N4020
18	21	22	190	1N3795A
20	125	25	40	1N2984A
22	4	30	2000	1N722
23.5	150	30	43	1N2049
24	4	30	1500	1N723A
30	15	45	1000	1N1782A
45	280	60	54	1N3329B
50	50	60	200	1N2996B
75	5	100	5000	1N1792A
91	140	125	243	1N3339
100	5	125	5000	1N1795
105	120	125	167	1N2839B
150	3	200	16,667	1N1890
200	65	250	769	1N2846B

2.3 CASCADED VOLTAGE REGULATOR

For better regulation than that afforded by the simple circuits of Fig. 2-1, two such circuits may be operated in cascade, as shown in Fig. 2-2. Here, the first stage (R1D1) of the circuit delivers a voltage, E2, having a certain percentage of regulation. This voltage then is applied to the second stage (R2D2), which smooths out the small variations in E2 and delivers a further regulated voltage, E3. In this circuit, I_{Z1} is the zener current of the first (input) diode, and I_{Z2} is the zener current of the second (output) diode.

The cascade circuit is designed in two steps: First, diode D2 and resistor R2 are determined (see Section 1.9, Chapter 1) for the desired final regulated output voltage (E3) for a convenient second-stage input voltage E2. Next, diode D1 and resistor R1 are determined (see Section 1.9, Chapter 1) for intermediate regulated voltage E2, for a convenient unregulated input voltage E1.

Very close stabilization of a dc voltage can sometimes be obtained with this simple circuit, provided a sufficiently high unregulated voltage E1 is available for the two-stage operation. Theoretically, there is no limit to the number of stages which may be cascaded in this manner, provided that the input voltage is high enough and that each stage is capable of supplying the current (loading) demands of the succeeding stage. However, these requirements them-

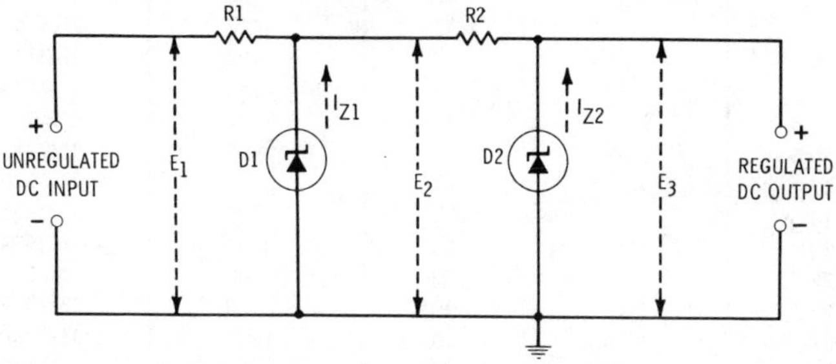

Fig. 2-2. Cascaded voltage regulator.

selves introduce a practical limit that will vary with individual circumstances.

2.4 DUAL-OUTPUT VOLTAGE REGULATOR

Two separate, regulated voltages may be obtained with the circuits shown in Fig. 2-3. The arrangement in Fig. 2-3A has the usual grounded-negative output; the similar arrangement in Fig. 2-3B, since it embodies a reversed input voltage and reversed diodes, provides grounded-positive output.

Each half of the circuit (i.e., R1D1 and R2D2) is designed separately according to the instructions given in Section 1.9, Chapter 1. From a single unregulated input voltage (E_i), the two regulated output voltages (E_{o1} and E_{o2}) may have the same or different voltage values, and so may the output currents.

The dual-output circuit is useful in certain instruments and devices in which separate operating voltages are desirable to prevent interaction between stages or to permit separate adjustment of operating or bias potentials.

Theoretically, there is no limit to the number of such simple regulator stages that may be operated in parallel with the unregulated input-voltage source. However, a practical limit is imposed by the

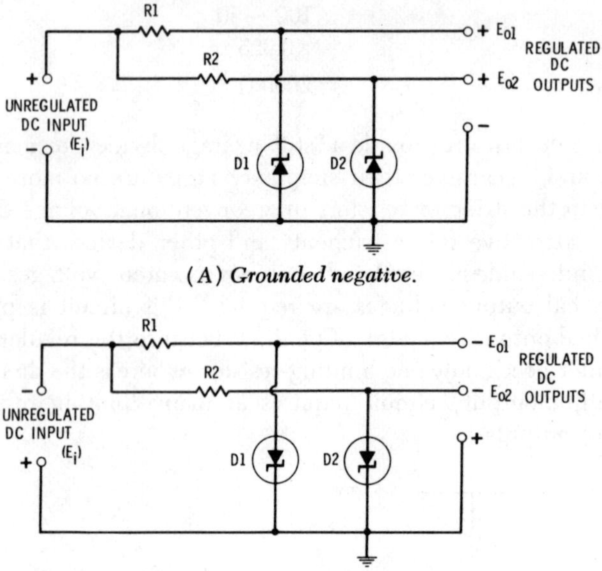

(*A*) *Grounded negative.*

(*B*) *Grounded positive.*

Fig. 2-3. Dual-output voltage regulator.

ability of that source to supply the total current required by the regulator circuits.

2.5 REGULATED VOLTAGE DIVIDER

A stable dc voltage divider may be made by substituting zener diodes for the resistor sections in a conventional voltage divider. By this means, each output voltage may be regulated automatically. Fig. 2-4 shows a typical circuit. Here, five selected 1N3020 diodes are employed to give dc output voltage in 10-volt steps from 10 V to 50 V. A similar combination of other diodes (selected for desired zener voltage V_Z and identical zener current I_Z) may be used for other output voltages.

Resistance R1 must be chosen to limit the current through the diode string to the I_Z value of the diodes in use, with respect to the unregulated dc input voltage. For the 1N3020 diode, the zener current is 25 mA; so, from the formula in Step 6, Section 1.9, R1 for a dc input of 100 V would be:

$$R1 = \frac{E_i - E_o}{I_Z}$$

$$= \frac{100 - 50}{0.025}$$
$$= 2000\ \Omega$$

Circuits of this sort provide a "stiff" voltage divider for most applications; and, because of their simplicity (they are no more complicated than the string of resistors in a conventional voltage divider), they are attractive for instruments and other devices that require several independent, regulated, but not isolated, voltages. Where nonidentical output voltages are required, this circuit is preferred to the dual-output regulator (Fig. 2-3), because the regulated voltage divider needs only one limiting resistor, whereas the dual-output (or multiple-output) circuit requires as many limiting resistors as there are outputs.

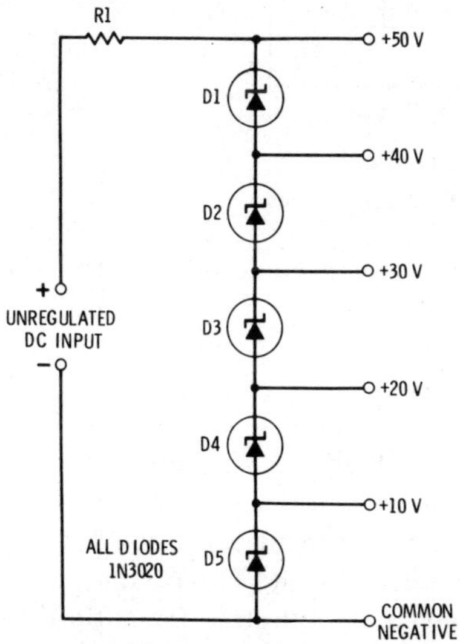

Fig. 2-4. Regulated voltage divider.

2.6 DUAL-POLARITY-OUTPUT REGULATED SUPPLY

Some electronic devices require simultaneous application of positive and negative dc voltages to different parts of their circuits. An example is the type of amplifier circuit that requires positive plate or collector voltage and negative grid or base voltage; another is the IC that requires positive and negative bias voltage. Such devices are powered by a supply unit having separate positive and negative outputs, or by separate supply units.

Fig. 2-5 shows a typical ac-line-operated supply delivering separate positive and negative output voltages $+E_o$ and $-E_o$. The positive regulator is comprised of zener diode D3 and limiting resistor R2, and the negative regulator of D4 and R4. Each of these regulators may be designed for the desired output voltage ($+E_o$ or $-E_o$), desired output currents, and available dc input voltage (E_{i1} or E_{i2}), according to the procedure explained in Section 1.9, Chapter 1.

While a single power transformer, T1, is shown here, separate half-wave rectifiers (D1 and D2) are used. The output of each rectifier is smoothed by a separate capacitor-input, RC filter (C1R1C2 for the positive voltage and C3R3C4 for the negative voltage). For simplicity, some designers dispense with limiting resistors R2 and R4 and select filter resistances R1 and R3 for combined filtering and diode-current-limiting functions. For increased filtering efficiency, resistors R1 and R3 may be replaced with iron-core filter chokes. When resistors alone are used, it must be remembered that the total resistance R1 + R2 (and R3 + R4 in the other half of the circuit) determines the current through diodes D3 and D4, respectively, not R2 and R4 alone.

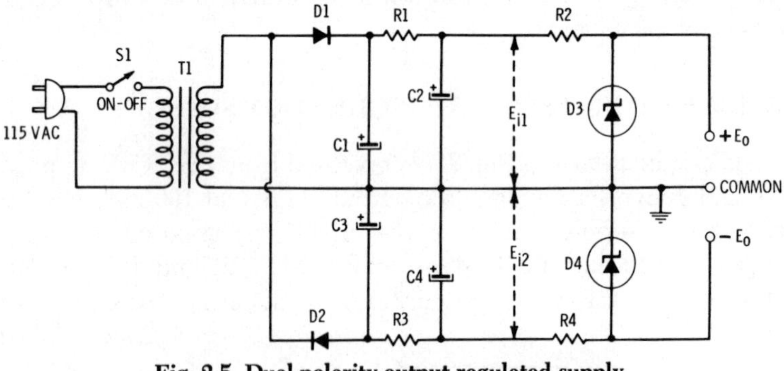

Fig. 2-5. Dual-polarity-output regulated supply.

2.7 LINE-OPERATED 150-V, LOW-CURRENT, REGULATED SUPPLY

Fig. 2-6 shows the circuit of a 150-V, 1-mA, voltage-regulated dc supply operated from the 115-V ac power line. A light-duty supply of this general type is useful for powering neon-bulb trigger and switching circuits and for similar experimental applications.

The power-supply portion is a simple half-wave circuit comprised of transformer T1, 1N92 rectifier D1, 62K load resistor R1, and 10-μF,

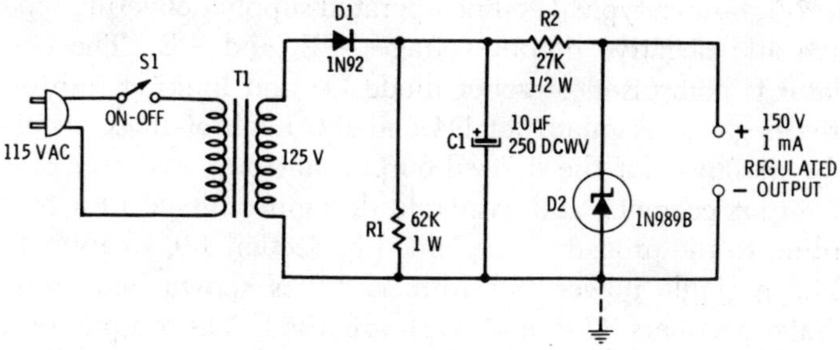

Fig. 2-6. Line-operated 150-volt, low-current, regulated supply.

250-V filter capacitor C1. The transformer (e.g., Stancor P-8624 with 12.6-V filament winding unused) delivers 125 V. The 10-μF filter capacitor (C1) charges to nearly 177 V, the peak value of the transformer secondary voltage.

The 1N989B zener diode, D2, regulates the dc output voltage to 150 V, with a 27,000-ohm, ½-watt, series, limiting resistor, R2. A similar circuit may be set up for other current and voltage values through appropriate choice of transformer secondary voltage and zener diode type, and by following the design procedure detailed in Section 1.9, Chapter 1.

2.8 LINE-OPERATED, 25-V REGULATED SUPPLY

The circuit shown in Fig. 2-7 is operated from the 115-V ac power line and delivers a regulated dc output of 25 V at 100 mA. The full-wave power-supply section of the circuit is composed of 30-volt-output transformer T1, bridge rectifier RECT1, and 100-μF, 50-V filter capacitor C1. A typical transformer is Stancor P-8614. The recti-

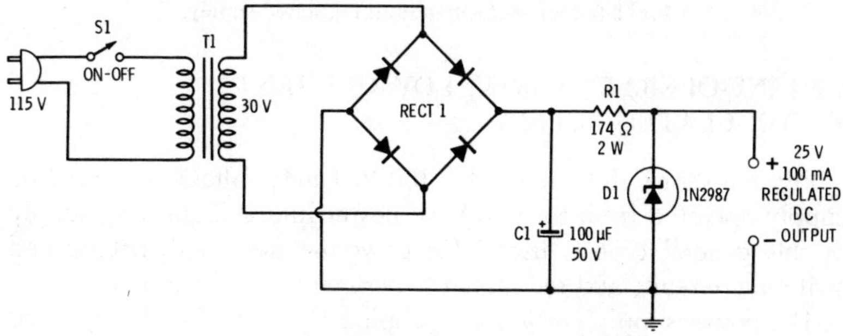

Fig. 2-7. Line-operated 25-volt regulated supply.

fier may be a standard full-wave bridge unit or may consist of four 1N4003 diodes connected as shown in Fig. 2-7.

Under no-load conditions, capacitor C1 charges to nearly 42.4 V, the peak value of the transformer secondary voltage. The 1N2987 zener diode, D1, regulates the dc output voltage to 25 V at 100 mA, with a 174-ohm series-limiting resistor, R1. (This resistor may need some adjustment with individual diodes.) A similar circuit may be set up for other voltage and current outputs through appropriate choice of transformer secondary voltage and zener diode type, and by following the design procedure given in Section 1.9, Chapter 1.

2.9 COMBINATION RECTIFIER AND REGULATOR

Fig. 2-8 shows a bridge composed of four zener diodes. This circuit both rectifies and regulates voltage. This dual action can greatly simplify a power supply when very precise voltage regulation is not required.

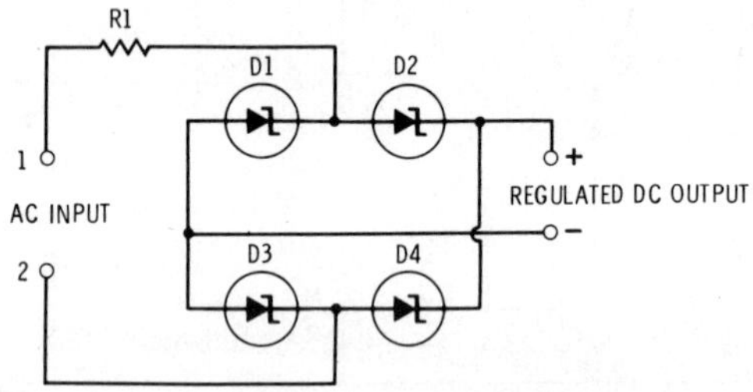

Fig. 2-8. Combination rectifier and regulator.

To understand how this circuit operates, note that on one ac half-cycle two of the diodes are forward-connected and the other two reverse-connected, and on the other half-cycle the connections of the diodes swap. Thus, when ac input terminal 1 is positive, D2 and D3 are forward-connected in series with the dc output terminals and rectify the ac, making the upper output terminal positive. At the same time, D1 and D4 are reverse-connected in shunt with the dc output and act as zener regulators. Conversely, when ac input terminal 1 is negative, D1 and D4 become the forward-connected rectifiers, and D2 and D3 the reverse-connected regulators.

The diodes must be chosen in type for the desired dc output voltage and current, with the zener voltage (V_Z) corresponding to the desired output voltage. They should be matched. (Matched quads

are commercially available.) Resistor R1 must be chosen, with respect to the diode specifications, to limit the zener current to the value specified by the manufacturer (see the design procedure detailed in Section 1.9, Chapter 1).

2.10 SERIES-TRANSISTOR VOLTAGE REGULATOR

Because the zener diode will deliver a constant dc voltage in a suitable circuit, it may be used as the dc reference source in a transistor-type regulated power supply. In tube-type power supplies, this function usually is performed by a gaseous regulator tube, but a series string of zener diodes may also be used.

In Fig. 2-9, D1 is a 1N3029B zener diode employed as the dc reference for transistor Q3. Transistors Q2 and Q3 constitute a dc

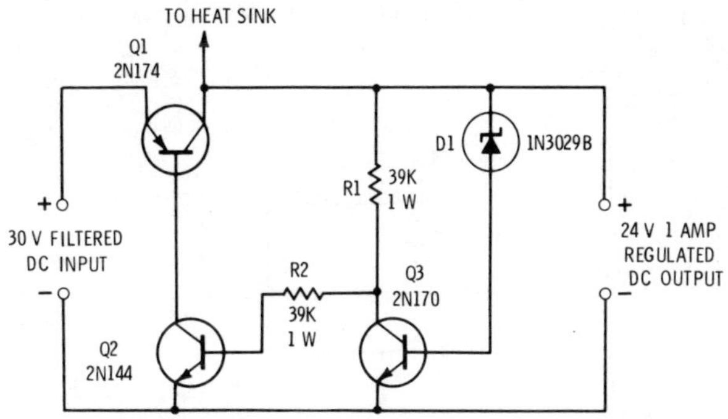

Fig. 2-9. A 24-volt series-transistor voltage regulator.

amplifier, which receives its input signal from diode D1 and delivers its output signal to transistor Q1. Transistor Q1 is connected between the + dc input terminal and the + dc output terminal, acting as a variable series resistor. Its resistance is controlled by current fed from the + dc output terminal back to the base of this transistor by dc amplifier Q2-Q3, an error-signal amplifier. Thus, when the output voltage rises, the resistance of Q1 is increased, and this reduces the output to the initial level. When the output voltage falls, the opposite action takes place.

This action is accomplished in the following manner: Normally D1 acts as an extremely high resistance, permitting almost no current to flow into the base of Q3. When the voltage at the + dc output terminal exceeds the breakdown voltage (BV), however, the diode

conducts, and this current is amplified by Q3 and Q2 in cascade, and is applied to Q1, altering the series resistance offered by Q1. The circuit delivers 24 volts at 1 ampere. Output regulation is 1%.

A similar arrangement may be employed for other regulated dc voltages and other current levels, through suitable choice of zener diode, transistors, and resistors.

2.11 LINE-OPERATED, 10-V SERIES-TRANSISTOR-REGULATED SUPPLY

Fig. 2-10 shows the circuit of a power-line-operated 10-V, 150-mA dc supply employing a zener-diode-referenced series-transistor volt-

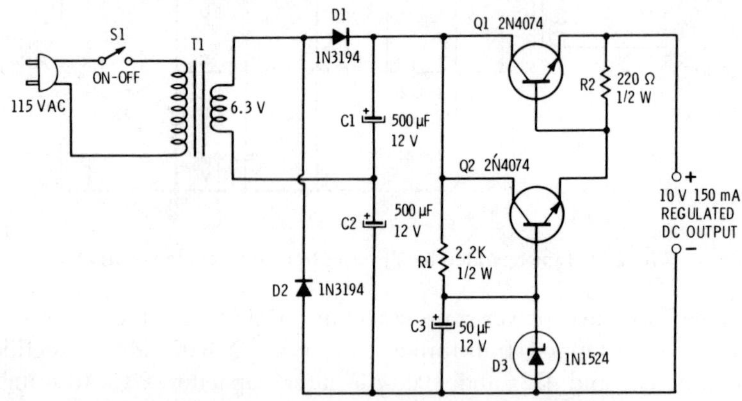

Fig. 2-10. Line-operated 10-volt series-transistor-regulated supply.

age regulator somewhat similar to the one described in the preceding Section 2.10. The power-supply section of this circuit is composed of 6.3-V transformer T1, two 1N3194 silicon rectifier diodes, D1 and D2, and two 500-μF capacitors, C1 and C2. This is a voltage doubler, with dc output voltage developed across C1 and C2 in series, equal approximately to 17.8 V (twice the peak value of the 6.3-V rms output of the transformer).

The voltage-regulator section of the circuit consists principally of the two 2N4074 transistors (Q1 and Q2) and the 1N1524 zener reference diode, D3. Here, Q1 is the controlled transistor, acting as a variable series resistor, and Q2 is the controlling transistor. The regulated output is 10 Vdc at 150 mA maximum. Some adjustment of the limiting resistance, R1, may be required for an individual zener diode (D3) and amplifier transistor (Q2).

2.12 LINE-OPERATED, 5-V SERIES-TRANSISTOR-REGULATED SUPPLY

A somewhat simpler line-operated, series-transistor-regulated supply than that described in the preceding section is shown in Fig. 2-11. This is a 5-V, 100-mA unit in which the series transistor (Q1) is directly controlled by the zener diode (D3); there is no intermediate amplifier.

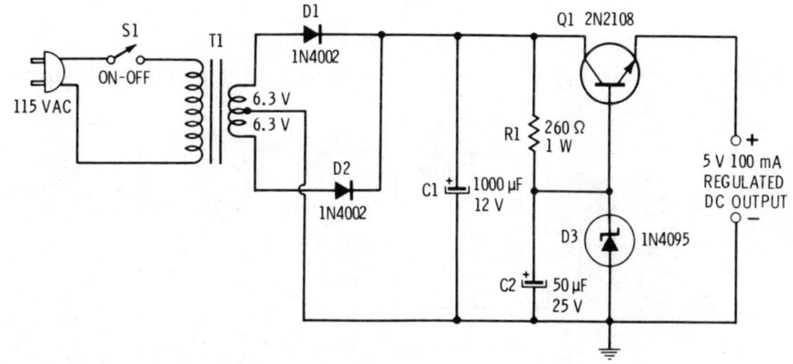

Fig. 2-11. Line-operated 5-volt series-transistor-regulated supply.

The full-wave power-supply section of this circuit consists of a 12-V center-tapped transformer, T1, two 1N4002 silicon rectifier diodes, D1 and D2, and 1000-μF filter capacitor C1. Additional filtering action is provided by the 50-μF capacitor, C2. Transformer T1 may be any convenient 12.6-V center-tapped filament or rectifier transformer.

The voltage-regulator section of the circuit consists of a 2N2108 transistor (Q1), acting as a variable series resistor, and the 1N4095 zener reference diode, D3. The regulated output is 5V at 100 mA maximum, but the same design may be used for higher outputs (up to several amperes at higher or lower voltage) by making suitable substitutions for T1, Q1, R1, and D3. Some adjustment of limiting resistance R1 may be required for an individual zener diode and transistor.

2.13 SHUNT-TRANSISTOR VOLTAGE REGULATOR

The preceding sections described voltage regulators based upon a zener-diode-referenced transistor acting as an automatic variable

resistor in series with the power supply and load. Figs. 2-12, 2-13, and 2-14, on the other hand, show circuits in which the controlled transistor is in parallel with the power-supply output terminals, i.e., in shunt with the load.

In Fig. 2-12, resistor R1 and the internal collector-to-emitter resistive path of the transistor (Q1) form a voltage divider for the unregulated input voltage, E_i. The output voltage, E_o, is taken across the collector and emitter terminals of the transistor, i.e., across the lower half of this voltage divider.

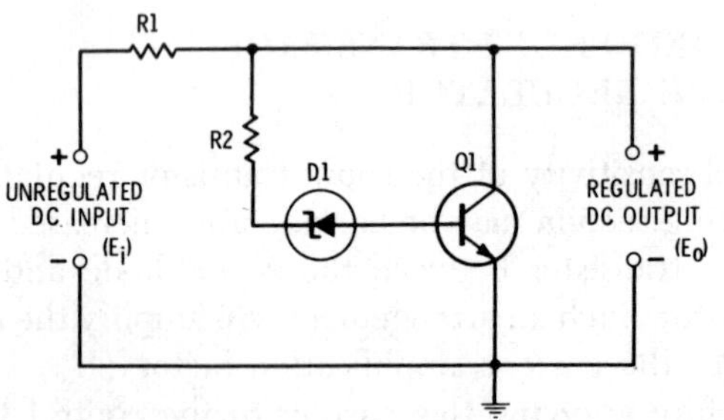

Fig. 2-12. Shunt-transistor voltage regulator (grounded negative).

The output voltage is sampled by zener diode D1 through the current-limiting resistor, R2; the resulting diode current (after the BV point of the diode has been passed) flows through the base-emitter input section of the transistor. Because of the inherent amplification of the transistor, this current is considerably increased in the collector-emitter circuit.

If the output voltage (E_o) decreases, owing to increased load-current drain, the diode current and, therefore, the transistor base-input current decrease. This causes less collector current to flow, the voltage drop across resistor R1 decreases proportionately, and the output voltage rises to restore its original value. The converse also is true when the load current decreases. The net result of this action is the automatic correction of output-voltage fluctuations, through the medium of the error current used to vary the resistance of the transistor member of the output-voltage divider.

The circuit in Fig. 2-12 has the usual grounded-negative output. In Fig. 2-13, however, the dc input terminals have been swapped in polarity; Q1 has been changed to a pnp transistor, and the zener diode has been reversed to provide grounded-positive output.

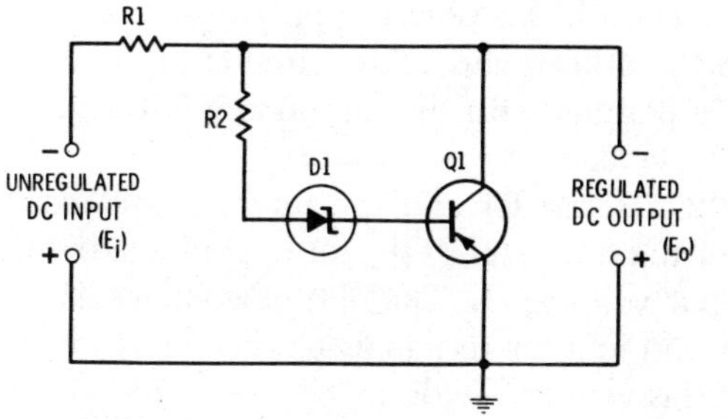

Fig. 2-13. Shunt-transistor voltage regulator (grounded positive).

2.14 CASCADED SHUNT-TRANSISTOR VOLTAGE REGULATOR

The control sensitivity of the shunt-transistor regulator described in the preceding section may be considerably increased by inserting a dc amplifier transistor between the zener diode and the output-control transistor. Such an arrangement will amplify the error current by as much as the current amplification factor (h_{fe}) of the added transistor, before applying this current to the control transistor. In this way, very small changes in output voltage may be compensated.

Fig. 2-14 shows the circuit of this latter type of shunt regulator. Here, the output transistor is Q1, and the amplifier transistor is Q2.

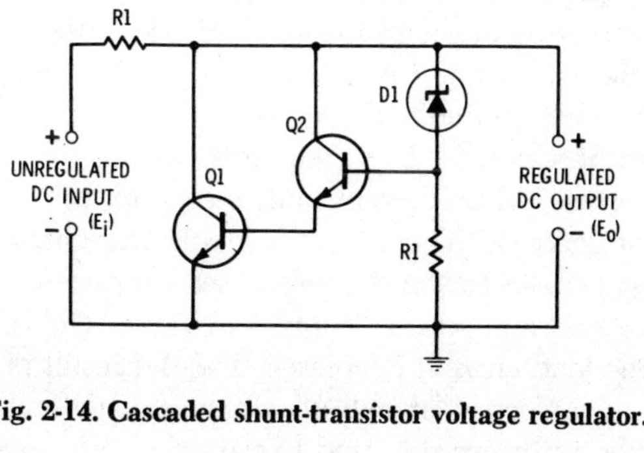

Fig. 2-14. Cascaded shunt-transistor voltage regulator.

As in the simpler circuit, Q1 forms an output voltage divider with resistor R1. Also as before, zener diode D1 samples the output voltage and, after the BV point of the diode is exceeded, passes current, this time into the base of the dc amplifier transistor, Q2. The amplified output current of Q2, then, is passed into the base-input circuit

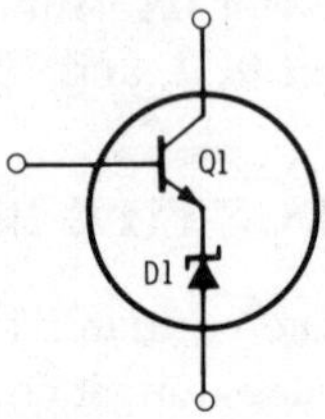

Fig. 2-15. Self-contained reference amplifier.

of transistor Q1, the latter altering its output resistance proportionately as a part of the output-voltage divider, R1Q1.

It is possible to combine zener diode D1 and transistor Q2 in a single unit called a *reference amplifier* (Fig. 2-15). Typical reference amplifiers are General Electric Types RA-1, RA-2, and RA-3.

2.15 ZENER DIODE IN TUBE-TYPE REGULATOR

For compactness, a zener diode may be used as the reference element in a tube-type voltage regulator. Fig. 2-16 shows this arrangement; here, zener diode D1 holds the cathode of amplifier-tube V2 at the voltage corresponding to the V_Z of the diode. Except for the inclusion of the zener diode, the circuit is entirely conventional: Tube V1 acts as a variable series resistor (comparable to transistor Q1 in Fig. 2-10), the output-error voltage is applied to the grid of dc amplifier tube V2, and the amplified output of V2 is applied to the grid of tube V1 to vary the latter's resistance.

Various zener diodes (listed with gaseous voltage-regulator tubes that they might replace in a tube-type regulator) include 1N1834A

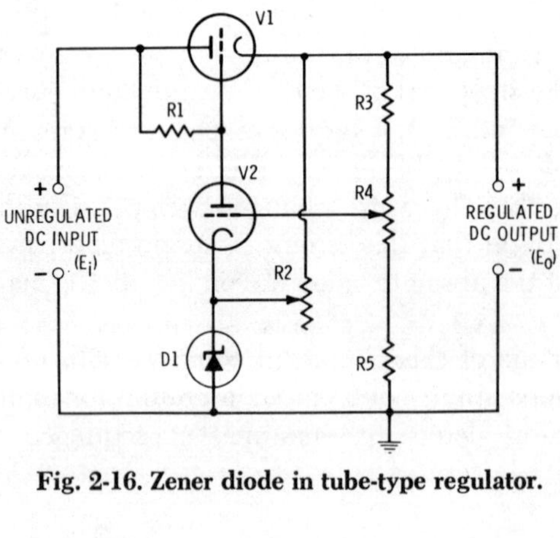

Fig. 2-16. Zener diode in tube-type regulator.

for the 75-volt 0A3 tube, 1N1836 for the 90-volt 0B3, 1N3006A for the 105-volt 0C3, and 1N3011A for the 150-volt 0D3.

2.16 BRIDGE-TYPE REFERENCE CIRCUIT

In voltage-regulator circuits in which the zener diode is the voltage-reference unit, increased sensitivity to error voltage may be obtained by incorporating the diode in the nonlinear bridge circuit shown in Figure 2-17A.

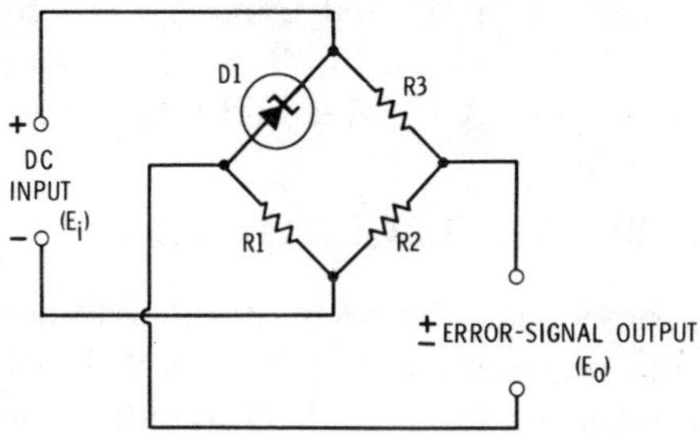

(A) Circuit.

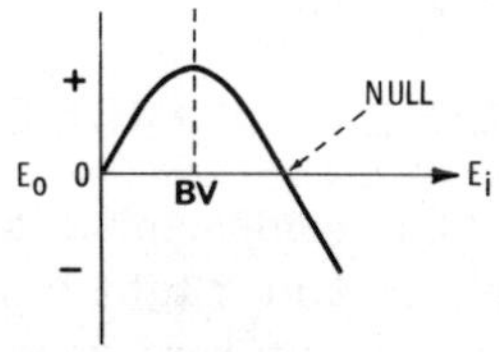

(B) Response.

Fig. 2-17. Bridge-type reference circuit.

Since the diode is, in effect, a voltage-dependent resistor (R_Z) the bridge is in a state of balance (i.e., output voltage E_o = zero) at only one value of input voltage E_i. At this point, $R_Z/R1 = R3/R2$, corresponding to the null point in Fig. 2-17B. At lower values of input voltage E_i, output voltage E_o is positive; at higher values of E_i, E_o is negative (again, see Fig. 2-17B). The bridge resistance values must be chosen so that at the desired power-supply output voltage, the bridge output voltage is zero.

Owing to the steepness of the diode conduction curve in the zener region (i.e., beyond BV), a very small change in error voltage (here, E_i) will unbalance the bridge and produce a large *change* in bridge output voltage E_o. The word "change" is emphasized here to thwart any misconception that this circuit is an amplifier: it is the change in voltage, not the absolute value of voltage, that is magnified by the circuit.

A disadvantage of this circuit, in common with other bridge circuits, is the impossibility of a common ground for input and output. But this natural defect has minimal consequence in many applications.

2.17 AUTOMOBILE VOLTAGE REGULATOR

Fig. 2-18 shows the circuit of an electronic voltage regulator for automobiles. In this arrangement, a zener diode (D2) senses the output voltage from battery B1, through the voltage divider R2-R3-R4.

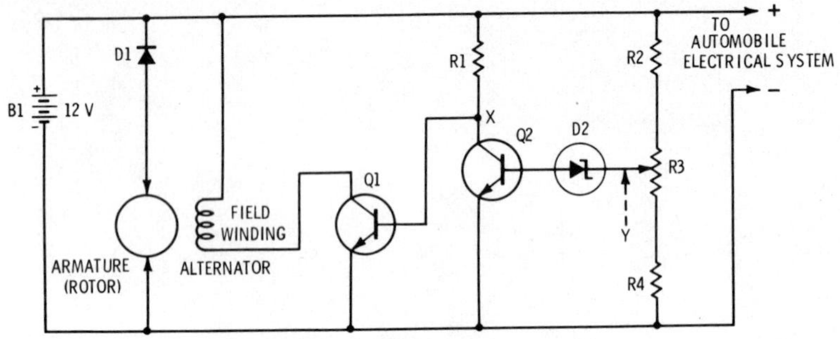

Fig. 2-18. Automobile voltage regulator.

In the voltage divider, potentiometer R3 is set so that zener current (I_Z) flows through diode D2 when the full-load output voltage of the circuit is correct (i.e., equal to the 12-V potential of the battery). Under these conditions, transistor Q2 amplifies the current it receives from D2, increasing the voltage drop across R1 and lowering the voltage at point X. This cuts off transistor Q1, because the latter receives little or no base-input signal from Q2 under these conditions. And, since Q1 acts as a variable resistor in series with the field winding of the alternator, no current flows through the field when Q2 is cut off, and the alternator voltage then is zero. The battery cannot discharge through the armature (rotor) of the alternator, because the rectifier (D1) is reverse-connected with respect to the battery.

When the output voltage falls, as when the loading is high or the battery charge depleted, the voltage at point Y drops proportionately. And, when this point-Y voltage falls below the breakdown voltage, BV, of the zener diode, transistor Q2 cuts off, reducing the voltage drop across R1 and increasing the voltage at point X. Transistor Q1 consequently conducts collector current through the field winding, and the alternator delivers output voltage, which is rectified by D1 and delivered to the car electrical system and to the battery, which it charges.

This action maintains a steady voltage at the output terminals, despite variations in system load and battery charge.

CHAPTER 3

AC Power Supply Applications

The zener diode has fewer ac than dc power-supply applications. Nevertheless, it enjoys some uses in which it significantly reduces circuit complexity and system size in ac voltage regulation.

This chapter describes several basic methods of ac voltage regulation with zener diodes. Some similarity will be discovered between these and the dc circuits of Chapter 2. In some instances, the reader may wish to combine one of these ac circuits with one of the dc circuits in the preceding chapter, for improved overall regulation.

3.1 BASIC REQUIREMENTS FOR AC REGULATION

The regulation of ac voltage requires that the positive and negative half-cycles both be automatically limited in amplitude. But, as has already been shown in Chapters 1 and 2, the basic zener regulator circuit operates on either positive or negative voltages, but not both simultaneously. This restriction demands that two zener diodes be employed in ac limiting, one for the positive half-cycle and the other for the negative half-cycle.

Offhand, it would appear sufficient to connect two diodes in parallel, anode to cathode; but this is unsatisfactory, because one diode would be forward-biased when the other was reverse-biased, and the forward one would conduct excessive forward current while the reverse one behaved normally as a zener diode. The correct pro-

cedure, on the other hand, is to connect the two diodes in series, cathode to cathode (or anode to anode). The reverse-biased diode then acts as a zener unit in series with the relatively small forward resistance of the forward-biased diode. When the ac cycle reverses, the forward diode becomes the reverse diode, and vice versa.

It is entirely feasible to connect two discrete diodes in series, and this, in fact, is often done. However, semiconductor manufacturers supply a self-contained unit—the *zener clipper* (Fig. 3-1)—which has the required two back-to-back junctions in a single unit, and use of this 2-terminal device greatly simplifies ac zener circuitry.

Fig. 3-1. Zener clipper.

3.2 BASIC AC VOLTAGE REGULATOR

Fig. 3-2A shows the circuit of the simplest full-wave, zener diode regulator. In this arrangement, two zener diodes (D1 and D2) are connected in series back-to-back, or a single zener clipper is used. As in the comparable dc circuit, a current-limiting resistor (R1) is connected in series with the diodes and the source of unregulated ac input voltage (E_i), and the regulated ac output voltage (E_o) is taken across the diode pair.

Operation takes place in the following manner: When input terminal 1 is positive, diode D1 is forward-biased and presents a low resistance to zener current I_Z flowing through reverse-biased diode D2. When resistance R1 is chosen correctly to limit I_Z to the recommended zener value, the positive half-cycle of ac output voltage (E_o) is limited to the zener voltage value by diode D2, as shown by the flattening of the positive half-cycle at V_{Z1} in Fig. 3-2B. When the input-voltage cycle reverses, input terminal 2 becomes positive, whereupon D2 becomes the low-resistance forward-biased diode and D1 the reverse-biased one. The latter then limits the negative half-cycle of ac output voltage, as shown by the flattening of the negative half-cycle at V_{Z2} in Fig. 3-2B.

If separate diodes are employed, their zener current (I_Z) ratings must be identical and must correspond to the desired load current. For symmetry in the output wave, their zener voltage (V_Z) ratings

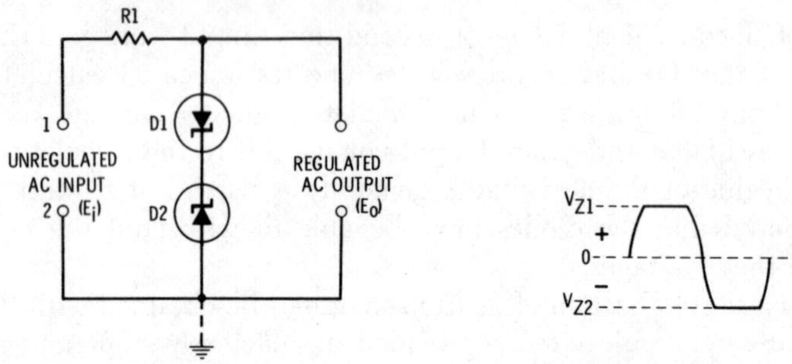

(A) Circuit. *(B) Output waveform.*

Fig. 3-2. Basic ac voltage regulator.

likewise must be identical. If a single zener clipper (e.g., Motorola Type 1M10ZZS) is employed, its I_Z rating should correspond to the maximum desired load current and its V_Z rating should correspond to the desired regulated output voltage.

The circuit may be designed in the same manner as described for dc circuits in Section 1.9, Chapter 1, except that the small forward resistance of the "standby" diode must be considered when calculating the series resistance, R1. In some cases, R1 will need some pruning for individual diodes or clippers.

It should be noted that the ouput of the ac voltage regulator is a clipped sine wave (that is, the waveform is squarish, being neither square nor fully sinusoidal) and may not be satisfactory for those power applications that require an undistorted sine wave.

3.3 CASCADED AC VOLTAGE REGULATOR

Two or more of the simple regulator sections from Fig. 3-2 may be connected in cascade for improved voltage regulation. Such an arrangement, embodying two stages, is shown in Fig. 3-3.

Here, the first stage (R1D1D2) of the circuit delivers an ac voltage, E2, having a certain percentage regulation. This voltage, then, is applied to the second stage (R2D3D4), which smooths out the small fluctuations remaining in E2 and delivers the further regulated voltage, E3. In this circuit, I_{Z1} is the zener current of the first (input) diodes, and I_{Z2} is the zener current of the second (output) diodes.

The ac cascade circuit, like its dc counterpart described in the last chapter, is designed in two steps: First, diodes D3 and D4 are selected (see Section 3.2) and resistance R2 calculated (see Section 1.9, Chapter 1) for the desired final regulated output voltage (E3) in

terms of a convenient value of second-stage input voltage (E2). Second, diodes D1 and D2 are selected and resistance R1 calculated (see Sections 1.9 and 3.2) for intermediate regulated voltage E2 in terms of available unregulated input voltage E1. In the calculations, the rms value of the ac voltages generally is used, but it must be remembered that the diodes must be able to withstand the peak value of these voltages.

Very close stabilization of ac can sometimes be obtained with this comparatively simple circuit, provided a sufficiently high unregulated voltage E1 is available for the two-stage operation. And the circuit may be simplified by employing two zener clippers in place of the four separate diodes in Fig. 3-3. Theoretically, there is no limit to the number of stages which may be cascaded in this manner, provided that the input voltage (E1) is high enough and that each stage is capable of supplying the current (loading) demands of the succeeding stage. However, as in the cascaded dc regulator, these requirements themselves introduce a practical limit that will vary with individual circumstances.

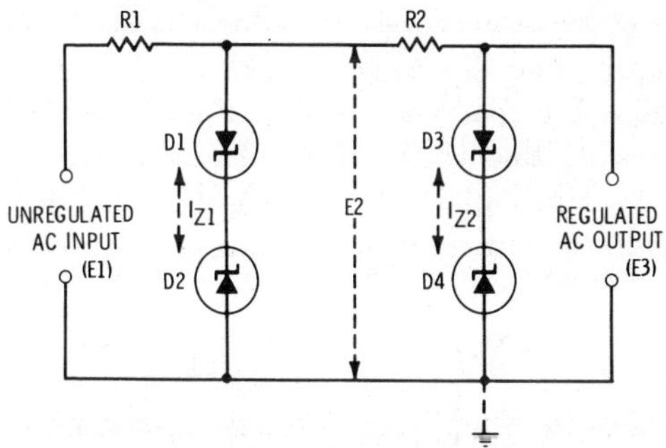

Fig. 3-3. Cascaded ac voltage regulator.

3.4 DUAL-OUTPUT AC VOLTAGE REGULATOR

Two or more simple ac voltage regulators, each with its own output terminals, may be operated from a single unregulated ac input voltage. A dual-output regulator of this type is shown in Fig. 3-4. Here, the first regulator is composed of current-limiting resistor R1 and back-to-back-connected zener diodes D1 and D2, and the second regulator is similarly composed of R2, D3, and D4. Separate regulated output voltages E_{o1} and E_{o2} are thus provided from the single unregulated ac input voltage, E_i.

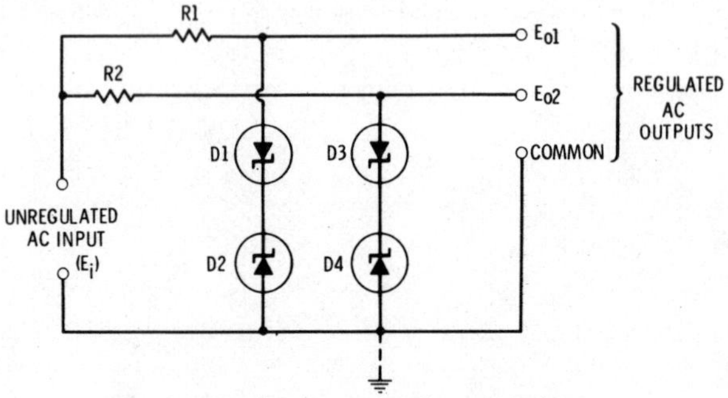

Fig. 3-4. Dual-output ac voltage regulator.

Each half of the circuit (R1D1D2 and R2D3D4) is designed separately according to the instructions given in Sections 1.9 (Chapter 1) and 3.2. From a single unregulated input voltage, (E_i), the two regulated output voltages (E_{o1} and E_{o2}) may have the same or different values, and so may the corresponding output currents.

Theoretically, there is no limit to the number of simple regulator stages that may be operated in parallel in this way from a single unregulated source. However, as in the dc counterpart of this circuit, described in Chapter 2, a practical limit is imposed by the ability of that source to supply the total current required by the regulator circuits.

CHAPTER 4

Miscellaneous Applications

The traditional association of the zener diode with voltage-regulated power supplies has somewhat obscured the other uses and potentialities of this device. This chapter describes a selection of such additional applications.

All uses of the zener diode must ultimately exploit the reverse-breakdown characteristics of this diode, and there are many electronic functions which may be simplified through use of breakdown diodes instead of more complicated devices or circuitry. However, not all such employments of the zener diode are completely practical. For that reason, some admittedly clever applications were excluded from this chapter.

Suggested applications, and the ones in this chapter are no exception, sometimes suggest others even if they themselves are not directly useful to the reader; in this way, a class of applications sometimes grows out of a single example. It is hoped that the applications described below may thus provide hints for further use, as well as prove useful in themselves.

4.1 DC VOLTAGE STANDARDS

Because a correctly operated zener diode maintains a constant dc voltage drop across itself, this device is naturally usable as a voltage standard. Voltage standards are used principally in potentiometric

measurements, the calibration of meters and other instruments, and the establishment of reference points for the operation of automatic control circuits. Conventional zener diodes may be employed, but for added stability, special *reference diodes* are commercially available with low temperature coefficients. Practical circuits for voltage standards are given in Figs. 4-2, 4-3, 4-4, and 4-5.

Fig. 4-1 shows the basic circuit of a dc voltage standard. This arrangement will be recognized as essentially the same as a voltage-

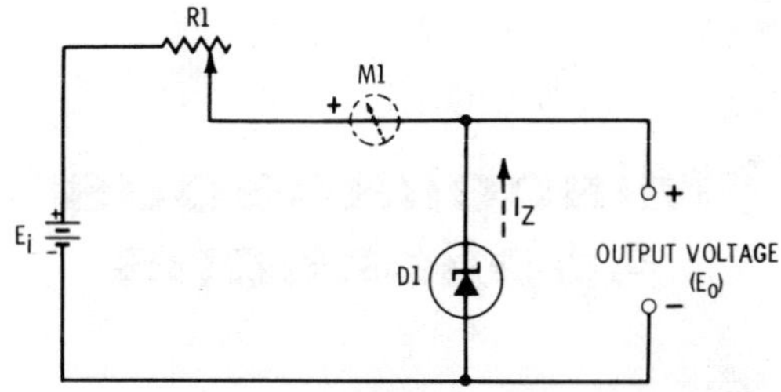

Fig. 4-1. Basic circuit of voltage standard.

regulator circuit. The difference is in the application: Specifically, a voltage standard is not usually required to furnish any appreciable output current; hence, external loading effects may usually be neglected when designing this circuit, since it is required only to compensate for the variations in input voltage. In Fig. 4-1, the input-voltage source (E_i) is shown as a battery, but it may also be an equivalent *well-filtered,* ac-operated dc power supply. Rheostat R1 is a current-limiting resistor that is set for a precise value of zener current, I_Z, through the zener diode, D1. As long as this value of current is maintained by periodic adjustment of R1 with respect to the deflection of dc milliammeter M1, the output voltage (E_o) will have the desired standard value, and the latter is equal to (or close to) the zener voltage, V_Z, of the diode.

The following design procedure is recommended:

1. Select a zener diode whose zener voltage rating, V_Z, is equal to the desired standard output voltage.
2. Note the value of zener current I_Z at which the diode manufacturer specifies V_Z.
3. Select a battery or power supply that will deliver a voltage, E_i, higher than V_Z.
4. Calculate the value of resistance R1 required for the correct I_Z

through the diode: $R1 = (E_i - V_Z)/I_Z$, where E_i and V_Z are in volts, and I_Z is in amperes. Include the internal resistance (R_m) of any milliammeter in the circuit in the total resistance required for rated I_Z.

It is good practice to split R1 between a variable and a fixed resistor (as in Figs. 4-2, 4-3, and 4-5) to prevent excessive diode current when the variable resistor is set to zero. Table 4-1 shows combinations of diode, resistor, and input voltage for 12 different standard voltages between 1.5 V and 30 V. Many other combinations, of course, are possible with the multitude of zener diodes now

Table 4-1. Diode/Resistor Combinations for Voltage Standards

Output Voltage (E_o, volts)	Zener Current (I_Z, mA)	Suggested Input Voltage (E_I, volts)	Total Series Resistance (R1, ohms)	Diode Type
1.5	30	3.0	50	1N3897
2.0	20	3.0	50	1N3898
3.9	50	6.0	42	1N1518
4.7	30	6.0	43	1N1508A
6.2	20	9.0	140	1N675
8.0	5.0	12	800	1N470
10	10	18	800	1N764
15	5.0	24	1800	1N666
20	6.2	30	1613	1N968
24	5.2	30	1154	1N970B
25	15	30	333	1N3951
30	4.0	45	3750	1N725

available. Column 3 of the table shows suggested input-voltage values. But any other available voltage, higher than the V_Z rating of the diode, may be used, provided the value of R1 in Column 4 is changed accordingly by means of the formula given in Step 4 above.

After the voltage standard is built, it should be adjusted initially by setting R1 for the exact value of output voltage E_o, as indicated by an accurate, high-resistance dc voltmeter (such as a vtvm or FETvm) connected temporarily to the output voltage terminals, and by noting the reading (I_Z) of milliammeter M_1. In subsequent use of the voltage standard, R1 always should be reset, if necessary, for this same milliammeter reading.

Fig. 4-2 shows a practical voltage-standard circuit with 10-V output. The 10-V drop appears across the 1N5240 zener diode, D1,

which has a zener current rating of 10 mA. The 24-V dc supply may be either a battery or a *well-filtered* ac-operated power unit. In this circuit, the 1000-ohm wirewound rheostat, R1, permits adjustment of the diode current for exactly 10 volts of output, while the 1000-ohm fixed resistor (R2) prevents excessive diode current when R1 is set to zero resistance. The internal resistance (R_m) of the 0-25 dc milliammeter (M1) is negligible with respect to R1 + R2; hence it does not enter into the calculation of the resistances. A similar voltage standard may be designed for a different output voltage by following the procedure outlined in the third paragraph of this section.

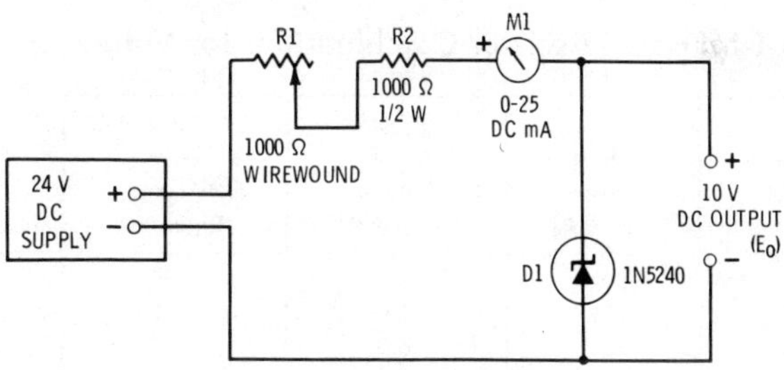

Fig. 4-2. A 10-v dc voltage standard.

Fig. 4-3 shows the circuit of a completely self-contained voltage standard based on a mercury battery, B1. Mercury batteries are noted for their low loss of voltage over a long period of use, but the added zener circuit compensates for even this small drift. Here, the 12.6-V mercury battery (B1, Mallory TR169) supplies the 1 mA of zener current required by the 1N3415 zener diode, D1. The required series resistance is 2600 ohms, supplied by the 2000-ohm wirewound rheostat (R1) and the 1600-ohm, ½-watt protective resistor (R2) in series. The internal resistance (R_m) of the 0-5 dc milliammeter (M1) is negligible, with respect to the 2600-ohm total, and does not enter into the resistance calculation. The output voltage, E_o, of the circuit will be 10 V, which is within the tolerance of the diode, when I_Z is adjusted to 1 mA by setting rheostat R1. However, the circuit should be initially set by adjusting R1 for exactly a 10-V output, as indicated by an accurate, high-resistance dc voltmeter (vtvm or FETvm) connected temporarily to the 10-V output terminals, and by noting the current indicated by the milliammeter. Whenever the circuit is used, R1, then, must be readjusted, when necessary, to restore that value of current. A similar self-contained voltage

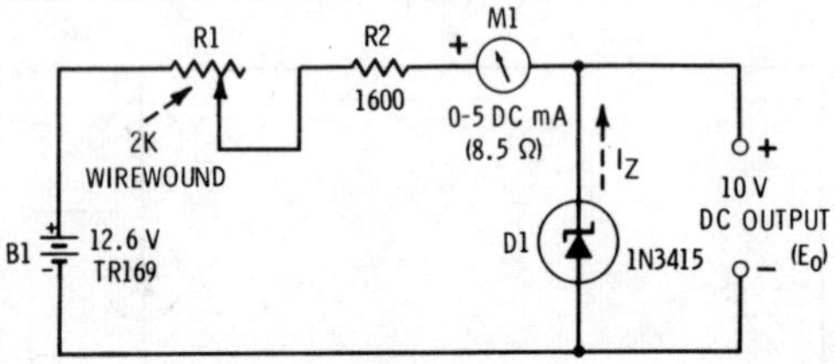

Fig. 4-3. Self-contained dc voltage standard.

standard may be designed for a different output voltage by selecting an appropriate battery and diode and by following the design procedure outlined in the third paragraph of this section.

Fig. 4-4 shows the circuit of a shunt-type dc voltage standard supplying five different output voltages. Each regulator leg in this circuit is seen to be identical to the basic circuit shown in Fig. 4-1, and all five legs are connected in parallel with the dc supply, which may be either a battery or a *well-filtered* ac-operated power supply. Each leg is designed for the desired output voltage by following the procedure outlined in the third paragraph of this section. While it is desirable to include a separate dc milliammeter (M1 to M5) in each leg of the circuit, a single meter may be switched in and out of the legs if this meter will handle all of the current values involved. There is no limit to the number of regulator legs which can be operated in a multioutput voltage standard, provided the dc supply can furnish the total current ($I_{Z1} + I_{Z2} + I_{Z3} + \ldots + I_{Zn}$) required by the circuit.

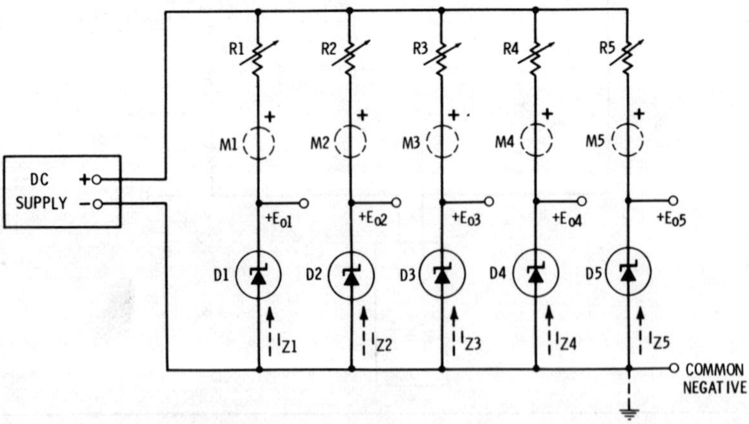

Fig. 4-4. Multiple-voltage dc voltage standard (shunt-type).

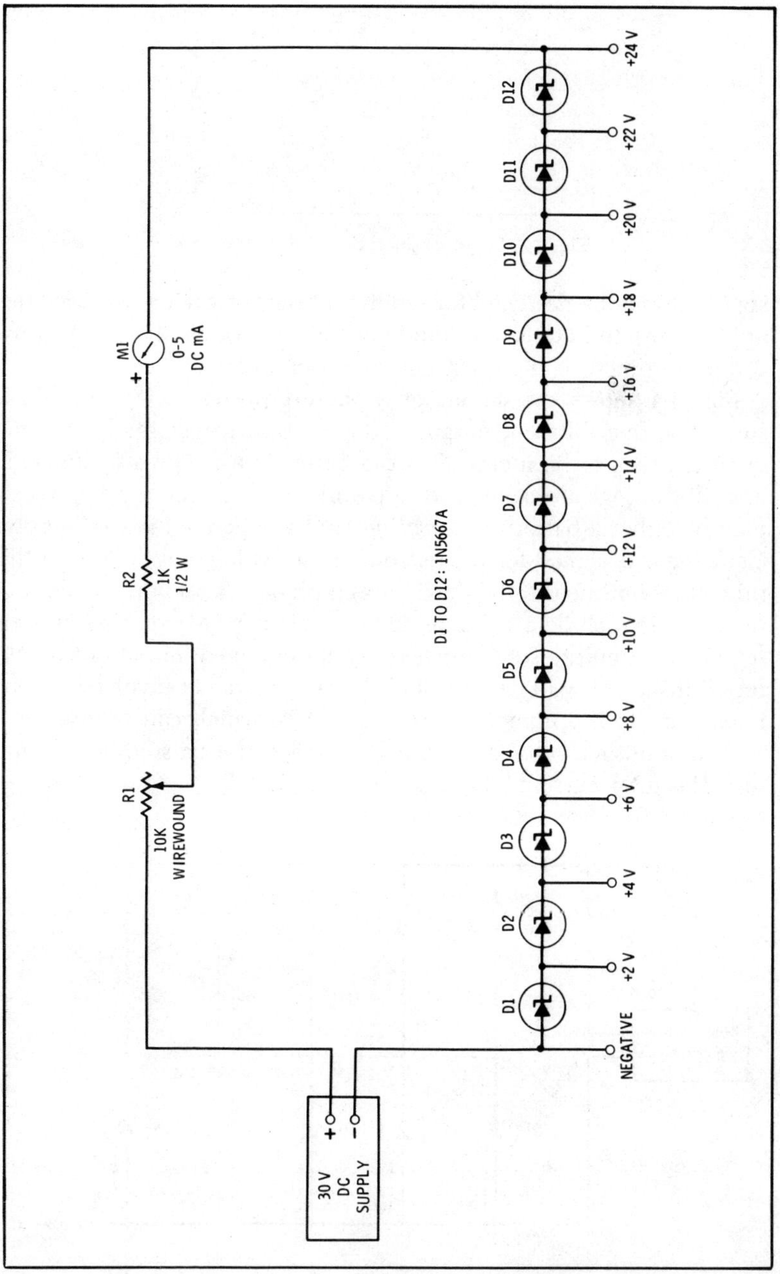

Fig. 4-5. Multiple-voltage dc voltage standard (series type).

Fig. 4-5 shows the circuit of a series-type dc voltage standard supplying 12 different output voltages in 2-volt steps from 2 V to 24 V. This arrangement is an adaptation of the zener-diode voltage divider described in Section 2.5, Chapter 2. In some instances, the series-type circuit will be preferred to the shunt-type circuit just described, since the series arrangement requires only one main leg. The series circuit employs 12 Type 1N5667A zener diodes, which, for best results, should be matched units. The 30-V supply may be either a battery or a *well-filtered* ac-operated power supply. With the 30-V supply, the required zener current of 1 mA is obtained through the series resistance composed of the 10,000-ohm wirewound rheostat (R1) and 1000-ohm, ½-watt, protective resistor (R2). The output voltages will have the values shown when R1 is set for 1 mA through the diode string, as indicated by the dc milliammeter, M1. A similar circuit may be set up for other voltages, provided all diodes used have identical I_Z ratings.

4.2 AUDIO-FREQUENCY CLIPPER

An automatic peak clipper for use in audio systems is shown in Fig. 4-6. This circuit will be recognized as similar to the basic ac-power-supply regulator described in Chapter 3. The zener-diode clipper circuit has the advantage over the conventional-diode clipper of not requiring the dc bias, which must be used with conventional diodes.

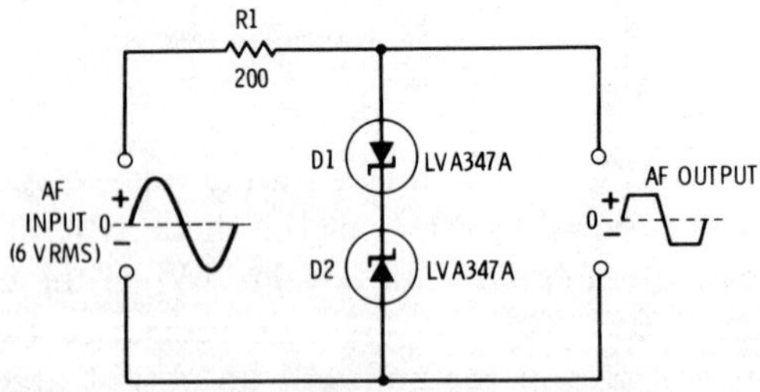

Fig. 4-6. Audio-frequency clipper.

The circuit, as shown, employs two Type LVA347A zener diodes (D1 and D2) connected in series, cathode-to-cathode, and a 200-ohm limiting resistor, R1. The diodes are of the avalanche type. The positive and negative peaks of the output signal are clipped and held to a constant amplitude when the input signal exceeds approximately

6 V rms. By its nature as an ac regulator, this circuit delivers a peak-to-peak output voltage that is somewhat lower than the peak-to-peak input voltage.

The clipper circuit may be modified to provide either a higher or lower clipping level by appropriate choice of diode and required series resistor. Observe that the output is a clipped sine wave, not a square wave.

4.3 ZENER-CONTROLLED RELAY

Below the BV point, the reverse resistance of a zener diode is high enough that the diode acts as an open switch; after the BV point, the resistance drops to such a low value that the diode acts as a closed switch. A zener diode, therefore, can function as a voltage-controlled switch in series with the coil of a dc relay. This arrangement is illustrated by Fig. 4-7.

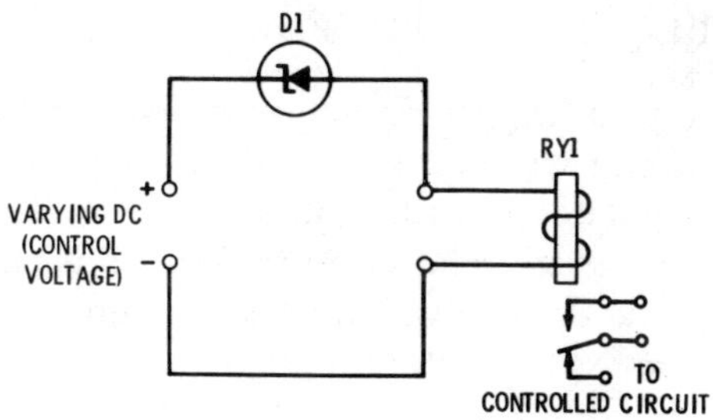

Fig. 4-7. Zener-controlled relay.

The diode (D1) is selected with respect to the dc resistance of the coil of relay RY1, so that the required value of pickup current through the coil is equal to the I_Z of the diode when the increasing dc control voltage reaches a critical value. When the coil resistance is equal to, or lower than, the static dc resistance (V_Z/I_Z) of the diode, the current rise in the coil will be abrupt at the switching point, and the relay will close quickly. Likewise, the fall of coil current will be equally abrupt as the control voltage decreases from the switching level, and the relay will open quickly.

The relay remains closed for all values of control voltage higher than that required for closure. Consequently, the maximum value of the input voltage must not be so high as to cause damage to the diode or relay.

4.4 ZENER-STABILIZED RELAY

The circuit described in the preceding section has the disadvantage that the relay coil is subjected to higher current than the pickup value as the input voltage increases significantly beyond the V_Z potential of the diode. This can result in damage to the relay, diode, or both, if the input voltage swing is excessive.

The circuit shown in Fig. 4-8 overcomes this fault by limiting the relay-coil voltage to the pickup value. This is accomplished by the regulator consisting of zener diode D1 and current-limiting resistor R1. The dc resistance of the relay coil constitutes the external load of the regulator circuit. The R1D1 circuit is designed in the same manner as any other zener-diode dc voltage regulator (see Section 1.9, Chapter 1), the diode being chosen so that its V_Z rating corresponds to the pickup voltage of the relay.

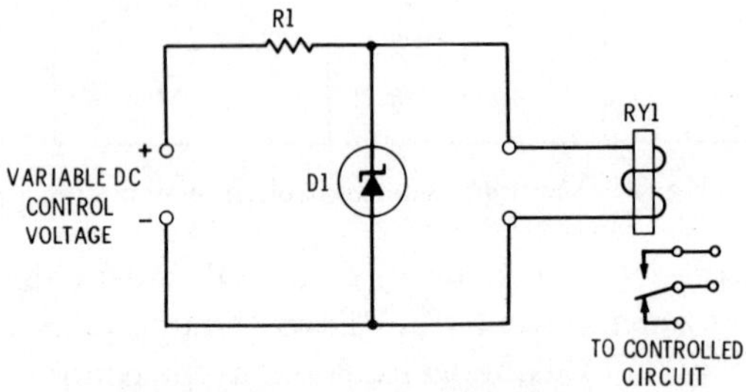

Fig. 4-8. Zener-stabilized relay.

Operation of the circuit takes place in the following manner: As the control voltage increases, the voltage applied to the relay coil increases proportionately. After the voltage passes the BV point of the diode, the relay voltage levels off at the V_Z potential of the diode (chosen to correspond to the relay-pickup voltage) and remains steady at this value as the control voltage rises still higher. Thus, the relay never receives a higher voltage than the rated pickup value, and its coil is protected.

4.5 MULTIPLE-RELAY CIRCUITS

The voltage-selective relay circuits described in Sections 4.3 and 4.4 are easily adapted to multiple circuits. Thus, several sections of either the zener-controlled circuit (Fig. 4-7) or the zener-stabilized circuit (Fig. 4-8) may be operated in parallel with the variable

control-voltage source to provide discrete relay operation at several control-voltage levels.

Fig. 4-9 shows a 3-section, zener-controlled relay circuit. Sequential operation of the relays is secured by proper choice of zener diodes D1, D2, and D3 with respect to the corresponding relay-coil resistance values and relay-coil currents I_{Z1}, I_{Z2}, and I_{Z3} (synonymous with the diode zener current values). When the control voltage

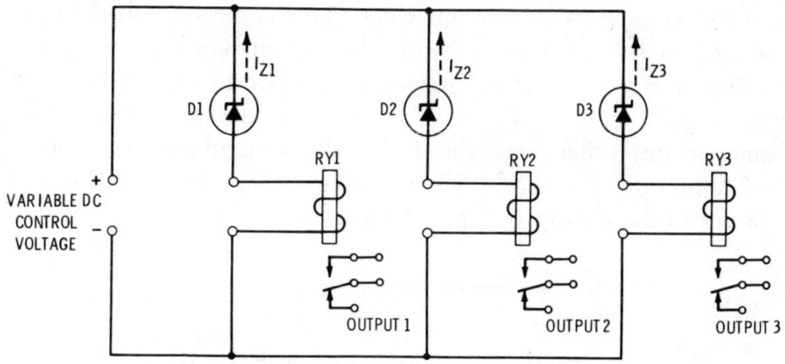

Fig. 4-9. Multiple, zener-controlled relay circuit.

reaches one predetermined level, diode D1 breaks down and the corresponding zener current (I_{Z1}) flows through relay RY1, closing the latter. Relay RY1 then remains closed as the control voltage rises higher. At a second predetermined value of control voltage, diode D2 breaks down and the corresponding zener current (I_{Z2}) flows through relay RY2, closing this second relay. Relay RY2 then remains closed, along with RY1. As the control voltage rises still higher, diode D3 breaks down and the corresponding zener current (I_{Z3}) flows through relay RY3, closing this third relay. Relay RY3 then remains closed, along with RY1 and RY2. As the control voltage subsequently decreases, the relays open in the reverse sequence (i.e., RY3 first, RY2 second, and RY1 third).

There is no limit to the number of legs such a circuit may have, so long as the control-voltage source can supply the total required current ($I_{Z1} + I_{Z2} + I_{Z3} + \ldots + I_{Zn}$). However, it must be kept in mind that all of the relays must be able to withstand the relatively high voltage required by the last relay in the network.

Fig. 4-10 shows a 3-section, zener-stabilized relay circuit. This circuit provides the same sequential pickup and dropout that are obtained with the circuit in Fig. 4-9. The difference is that each

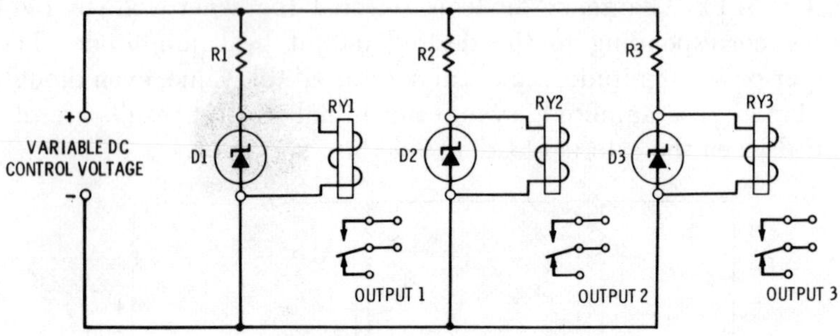

Fig. 4-10. Multiple, zener-stabilized relay circuit.

relay experiences a coil voltage that cannot exceed its required pickup value; in Fig. 4-9 each relay eventually "sees" the highest voltage applied to the last relay in the network. Each leg in Fig. 4-10 is similar to the basic leg shown in Fig. 4-8 and is designed in the same way.

Like the circuit given in Fig. 4-9, there is no limit to the number of legs which the multiple zener-stabilized relay circuit (Fig. 4-10) can have, so long as the control-voltage source can supply the total current required by the regulator/relay legs.

4.6 PULSE LIMITER

The simple, two-component circuit shown in Fig. 4-11 receives pulses of varying height and delivers pulses of constant height. The peak amplitude of the ouput pulse corresponds to the zener voltage (V_Z) of the diode, D1.

The circuit is basically a dc voltage regulator and requires the same design procedure as that for the dc voltage standard (see

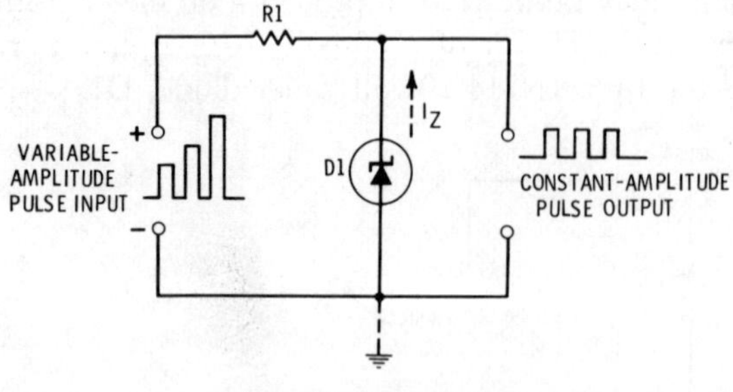

Fig. 4-11. Pulse limiter.

Section 4.1). The zener diode is selected for zener voltage (V_Z) rating corresponding to the desired output peak amplitude. The output-pulse amplitude, then, will not exceed this value, even though the input-pulse amplitude swings much higher. Best results are obtained when the external loading is light.

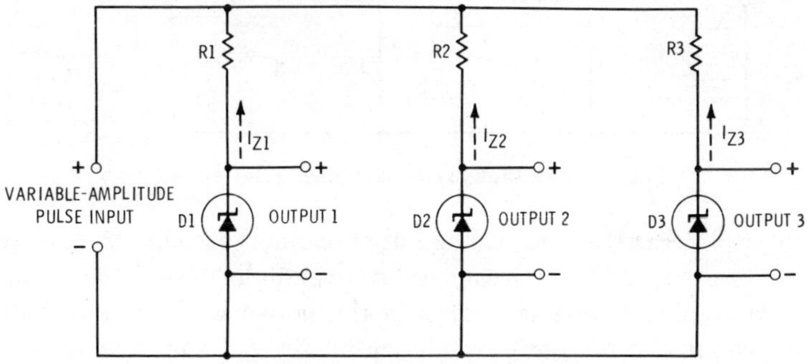

Fig. 4-12. Multiple pulse limiter.

Several similar circuits with diodes having different V_Z ratings may be operated in parallel with the input-pulse source to provide output pulses at selected constant amplitudes and identical pulse repetition rate. A multiple circuit of that type, containing three limiter legs, is shown in Fig. 4-12. The maximum number of legs which such a circuit can have depends upon the ability of the input-pulse source to supply the total required current ($I_{Z1} + I_{Z2} + I_{Z3} + \ldots + I_{Zn}$) and the extent to which the added legs distort the pulse shape.

4.7 ZENER-DIODE METER PROTECTOR

The sharp breakdown of a zener diode from its previous high-resistance state may be utilized to protect a dc meter from burnout. Fig. 4-13 shows the circuit of a 50-volt, 20,000 ohms-per-volt voltmeter protected by a 1N714 10-volt zener diode, D1.

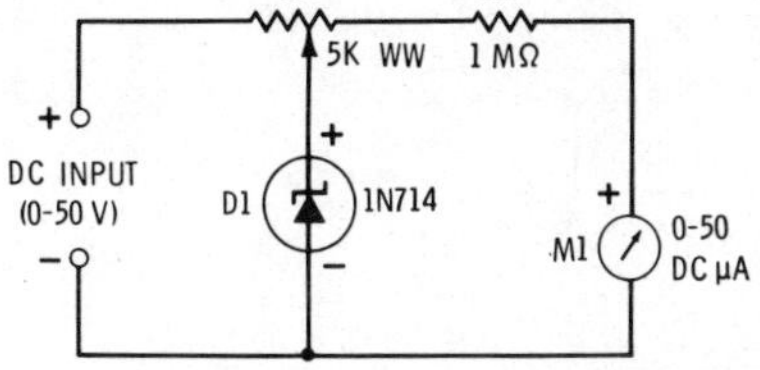

Fig. 4-13. Zener diode meter protector.

Here, the indicating meter (M1) is a 0-50 dc microammeter, whose scale reads 0-50 volts without requiring mental addition of ciphers. Resistor R2 is the principal multiplier for the voltmeter. Potentiometer R1 is set to the position at which, with 51 volts applied to the dc input terminals, the diode begins to conduct in its breakdown region. After this adjustment, the diode acts as an extremely high resistance so long as the input voltage is between zero and 50 volts; no voltage drop due to diode current is produced across R1, and the meter accuracy is not affected. As soon as the input exceeds 50 volts, however, the diode conducts, and the resulting voltage drop across R1 prevents overswing of the meter.

This is a fast-acting system, much more rapid in its action than a protective relay can be and one which will protect any sensitive meter in a similar circuit.

4.8 ZENER DIODE IN TRANSISTOR BIAS NETWORK

For maximum stability, a stiff voltage divider, operated from the main power supply, is used to furnish dc base bias to a common-emitter-connected transistor. If a zener diode biased to conduct in its breakdown region is used as the bottom section of the divider, the regulating property of this diode will provide a stable bias voltage.

In Fig. 4-14, resistor R1 and zener diode D1 form the bias voltage divider. Resistance R1 must be chosen with respect to the power-supply voltage (E) so that rated zener current (I_Z) flows through the diode. The zener voltage (V_Z) of the diode must be the same as the desired base-bias voltage. Thus, in the i-f amplifier in Fig. 4-14, D1 is a Type 1N3897 diode (zener voltage = 1.5 V), R1 is 375

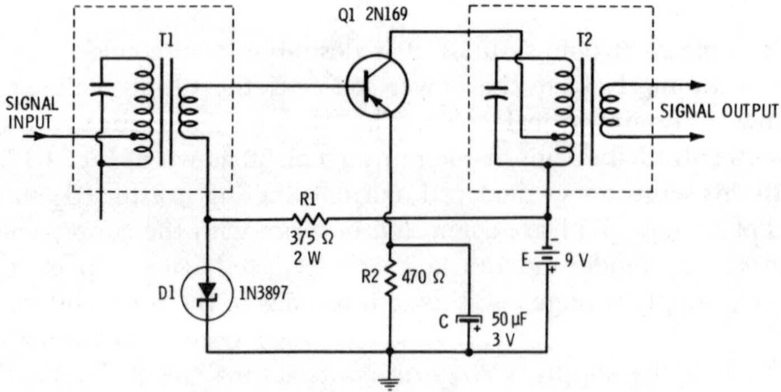

Fig. 4-14. Zener diode in transistor bias network.

ohms, 2 watts, and E is 9 volts. Because of its low dynamic impedance (5 ohms), the diode does not need a bypass capacitor.

The cost of the increased stability provided by this bias scheme is the relatively high current which must flow through the diode (30 mA for the 1N3897). But where high stability is imperative, the attendant expenditure is small and the method is simple.

4.9 ZENER DIODE AS TUBE CATHODE RESISTOR

Automatic bias in a tube-type amplifier or oscillator may be supplied to advantage by a zener diode, as shown in Fig. 4-15. From the many available types, a diode should be selected that has the desired bias voltage as its V_Z value, and for which the maximum dc cathode current of the tube equals the zener current (I_Z) of the diode.

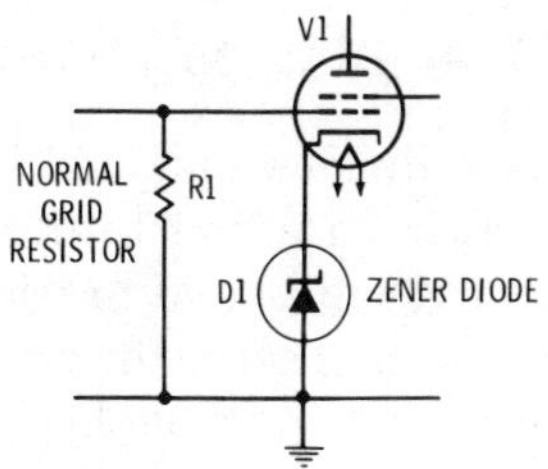

Fig. 4-15. Zener diode as tube cathode resistor.

The peculiarities of this scheme are: (1) the tube current can vary considerably without materially changing the bias voltage developed across the diode, and (2) the dynamic impedance of the diode is so low that no cathode bypass capacitor is required.

4.10 UNDERVOLTAGE PROTECTOR

In some electronic systems, it is desirable to disconnect the load device promptly from the power source if the supply voltage falls below a certain critical value. This is done automatically by the zener-controlled circuit-breaker arrangement shown in Fig. 4-16.

In this setup, zener diode D1, current-limiting resistor R1, and the coil of dc relay RY1 are connected in series with the supply-voltage source. The diode and resistance are selected with respect to the normal supply voltage and the dc resistance of the relay coil, so that the diode breaks down and conducts zener current I_Z through the coil when the supply voltage is at its normal level. Under these conditions, the relay is closed and connects the load to the source.

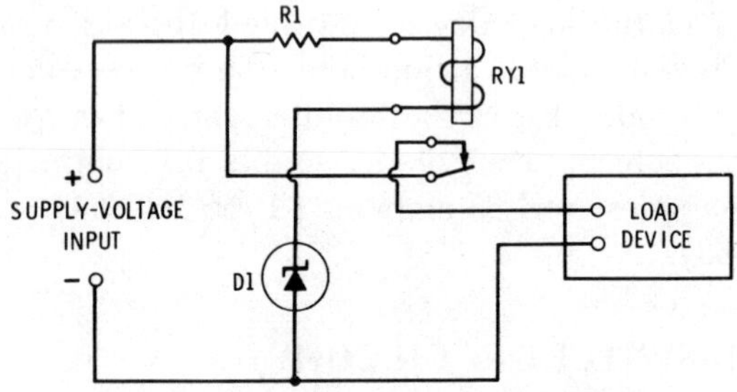

Fig. 4-16. Undervoltage circuit breaker.

Whenever the supply voltage falls below the threshold determined by the zener circuit, the coil current drops abruptly and drastically, and the relay opens quickly, disconnecting the load from the source.

4.11 OVERVOLTAGE PROTECTOR

Electrical devices often need protection against unexpected surges in supply voltage. Fig. 4-17 shows an arrangement for providing this protection automatically without disconnecting the device from the power source.

In this setup, a simple voltage regulator—consisting of zener diode D1 and current-limiting resistor R1—is inserted between the voltage supply and the device. The diode is selected for a V_Z rating equal to the normal operating voltage of the device, and for an I_Z rating high enough to accommodate the operating current of the load device; resistance R1 is calculated with these factors in mind. The source voltage (E_i) must be somewhat higher than the required device voltage (E_o), in accordance with zener-diode voltage regulator design (see Section 1.9, Chapter 1).

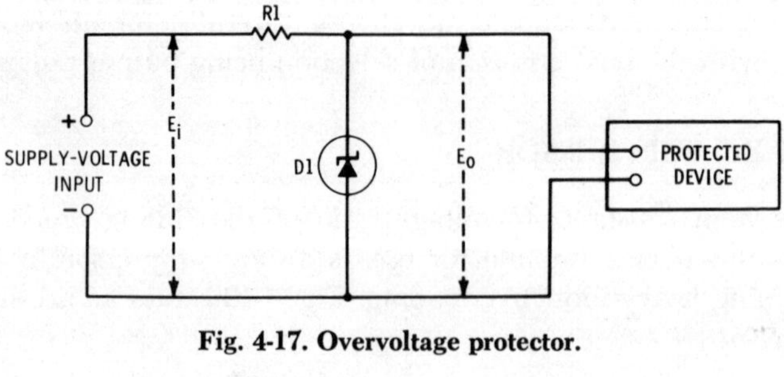

Fig. 4-17. Overvoltage protector.

When supply voltage E_i rises higher than the normal operating voltage (E_o) of the load device, the diode breaks down, and the resulting voltage-regulator action holds the device voltage constant at the required value E_o. High absolute values of current and voltage are obtainable in zener diodes usable for control purposes of this type. Examples are 1.85 amperes (Type 1N2804) and 560 volts (Type 1N1953).

4.12 ZENER SWITCHING CIRCUIT

In a diode-type pulse switching circuit, a zener diode does not introduce the pronounced carrier-storage effects that sometimes distort the pulse when switching from positive to negative with conventional diodes. Fig. 4-18 shows the basic arrangement of the zener-type pulse switch.

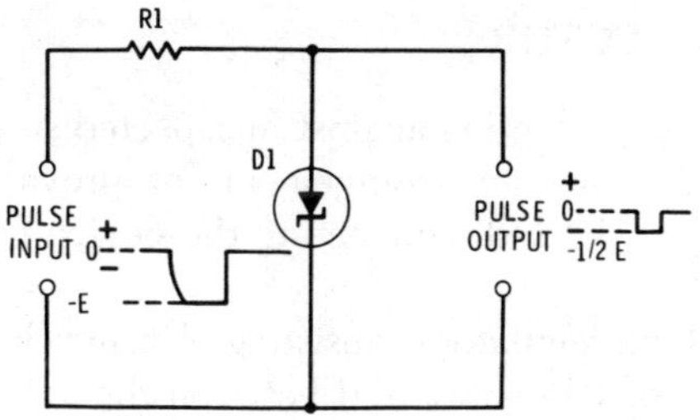

Fig. 4-18. Zener switching circuit.

While the output pulse delivered by the zener circuit is lower in amplitude than the input pulse, owing to the inherent voltage-divider action of the circuit, the output pulse is sharpened—as shown in the wave patterns in Fig. 4-18—when the input pulse exhibits rounding due to loss of high-frequency components. This is an additional advantage of the zener circuit.

For a particular pulse repetition rate and input-pulse peak amplitude, resistance R1 should be chosen carefully for an individual diode, with the final criterion of selection being output-pulse shape.

4.13 ARC SUPPRESSOR

The high resistance-to-voltage ratio of the "idling" zener diode makes this device suitable for use as an arc suppressor in circuits containing heavy inductive loading. Fig. 4-19 shows an arc-suppressor circuit.

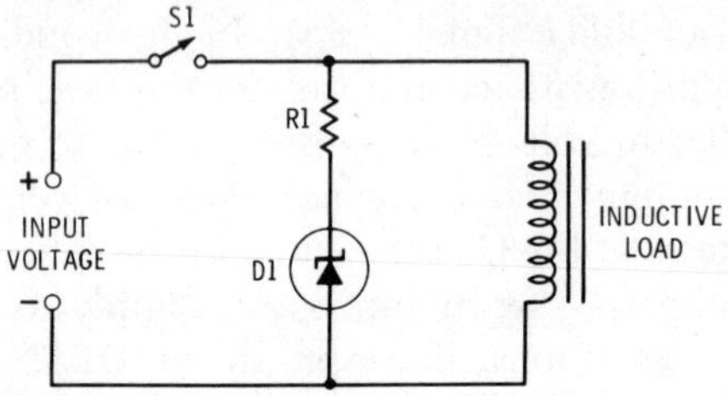

Fig. 4-19. Zener diode as arc suppressor.

Without the protective circuit (R1D1), the high back voltage developed by the collapsing magnetic field of the device causes an arc to form between the switch contacts when switch S1 is opened after the load device has been conducting current. With the suppressor circuit in place, however, the heavily conducting diode short-circuits the high voltage, thus protecting the switch contacts.

Because of its reverse polarity with respect to the input voltage, the diode acts as an extremely high resistance when the switch is closed and the load is conducting a steady direct current; under these conditions, the suppressor circuit offers virtually no load to the power supply.

4.14 INTERSTAGE COUPLING DEVICE

The insertion of a suitably chosen zener diode between the stages of a direct-coupled amplifier greatly simplifies the dc supply requirements of such an amplifier. Fig. 4-20 shows the skeleton circuit of a zener-coupled amplifier; for simplicity, the bias networks and dc supply have been omitted.

When the diode is properly chosen with respect to its V_Z rating and the required dc voltages of the circuit, it operates below its BV

Fig. 4-20. Zener diode as interstage coupling device.

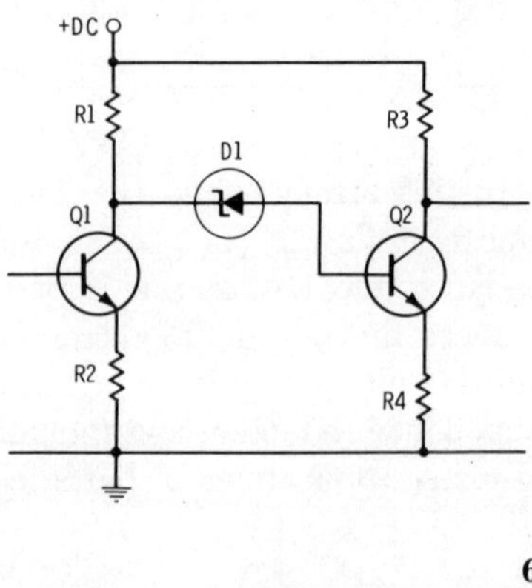

point when the amplifier signal is zero. In this condition, the diode acts as a very high resistance and isolates the base of transistor Q2 from the dc collector voltage of transistor Q1. (In the conventional direct-coupled amplifier circuit, which does not contain the diode, this voltage appears at the Q2 base and must be cancelled, a requirement that complicates the circuit, power supply, or both.) Under maximum-signal conditions, however, diode D1 is driven into its zener breakdown region and acts as a low resistance to couple the signal from Q1 through Q2. By this action, the diode preserves the advantages of direct interstage coupling (compactness, simplicity, good frequency response, etc.) while providing dc isolation similar to that afforded by a coupling capacitor, but without the disadvantages of the capacitor. The low dynamic impedance of the zener diode is very desirable in the coupling of ac signals.

Although a transistor amplifier is shown in Fig. 4-20, zener-diode coupling may be employed in tube-type amplifiers, as well.

4.15 BATTERY ISOLATOR

Fig. 4-21 shows the circuit of a transistor-type battery isolator in which a Type 1N706A, 6.1-V zener diode (D1) is the reference unit. This circuit is seen to be of the general type of dc voltage regulator described in Chapter 2.

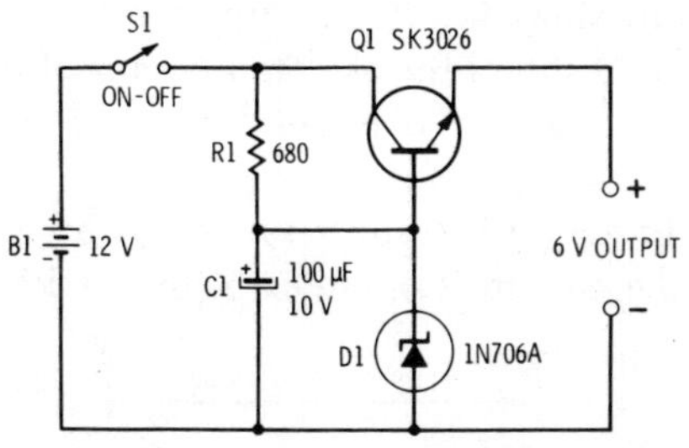

Fig. 4-21. Battery isolator.

In this arrangement, the Type SK3026 transistor operates as an emitter follower (i.e., as an isolator stage having high input impedance and low output impedance), and accordingly it in effect reduces the output impedance of battery B1, as well as provides output-voltage regulation against the drift of the aging battery. The impedance reduction and the voltage regulation both are invaluable in many applications of batteries to solid-state circuits.

4.16 ZENER DIODE IN FIXED-BIAS SUPPLY

A zener diode makes a simple, but effective, regulator for the fixed bias of an amplifier or oscillator. Fig. 4-22 shows such an arrangement for holding constant the fixed negative bias of a tube stage. The same arrangement may be employed for a transistor stage, and positive—instead of negative—bias may be regulated simply by reversing diode D1.

The diode is selected for a V_Z rating equal to the desired bias voltage, and R1 is calculated in terms of the input bias-supply voltage (E_c) and the zener current (I_Z) of the diode. (See Section 1.9, Chapter 1.) Here, R2 is the normal grid resistor of the stage. Because of the low dynamic impedance of the diode, no bypass capacitor is ordinarily required. However, in rf amplifiers and oscillators, a radio-frequency choke (rfc) will be required, as shown by the dotted symbol.

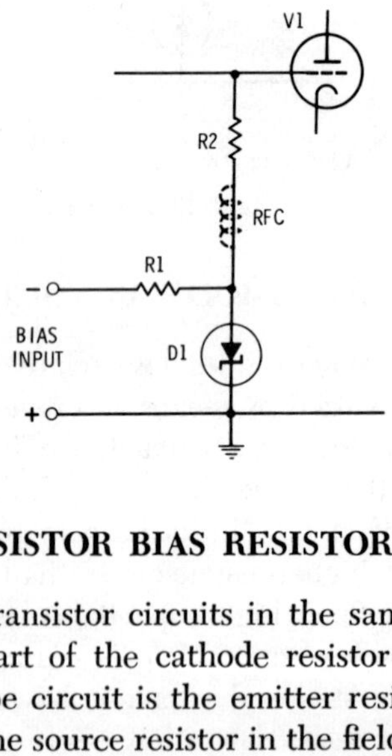

Fig. 4-22. Zener diode in fixed-bias supply.

4.17 ZENER DIODE AS TRANSISTOR BIAS RESISTOR

Automatic bias is obtained in transistor circuits in the same way as in tube circuits. The counterpart of the cathode resistor which supplies automatic bias in the tube circuit is the emitter resistor in the bipolar transistor circuit and the source resistor in the field-effect transistor circuit. In the transistor circuit, the bias resistor may be replaced with a zener diode, just as was done in the tube circuit in Section 4.9.

Fig. 4-23 shows two uses of the zener diode as an automatic bias resistor: in a common-emitter bipolar transistor circuit in 4-23A and in a common-source field-effect transistor circuit in 4-23B. From the

many available types, a diode should be selected that has the desired bias voltage as its V_Z value, and for which the maximum dc emitter current (bipolar transistor) or maximum dc source current (field-effect transistor) equals the zener current (I_Z) of the diode.

As in the tube circuit, the features of this scheme are: (1) the pertinent transistor current can vary considerably without materially changing the bias voltage developed across the diode, and (2) the dynamic impedance of the diode is so low that no emitter (or source) bypass capacitor is required.

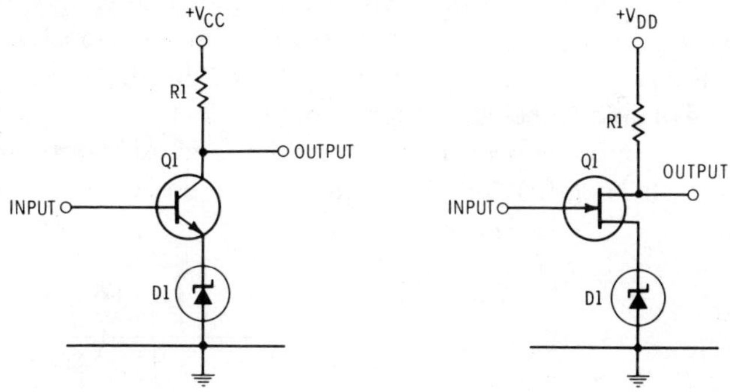

(A) *Bipolar transistor.* (B) *Field-effect transistor.*

Fig. 4-23. Zener diode as transistor bias resistor.

4.18 ZENER DIODE AS VOLTAGE-VARIABLE CAPACITOR

Most zener diodes have relatively high junction capacitance which can be varied by means of a reverse dc voltage. The higher the voltage, the lower the capacitance, but the voltage must always be well under the BV value to keep the diode from drawing zener current. This effect may be utilized to provide a voltage-variable capacitor having higher capacitance than that offered by regular varactor diodes. Such high capacitance is invaluable for tuning af circuits for which a conventional variable capacitor would be gigantic. Unlike the standard varactor, however, the zener diode used in this manner does draw a few microamperes (around the I_R value of the diode) from the dc control-voltage source.

Fig. 4-24 shows the circuit of a zener-type voltage-variable capacitor. The adjustable dc control voltage is applied to terminals 3 and 4, and may be obtained from an adjustable ac-operated dc power supply or from a battery-and-potentiometer combination. The junction capacitance of zener diode D1 varies inversely with this volt-

age and is "seen" at terminals 1 and 2. Isolation is provided by the 1-megohm resistor, R1, to block leakage of ac into the dc supply (a much bulkier af choke might be used instead of the resistor).

The zero-voltage capacitance and the extent to which it is varied by a given dc voltage change vary with the individual diode, since zener diodes are not intentionally processed for any certain guaranteed capacitance. Using a Type 1N1604 zener diode in this circuit, the author obtained a capacitance change from 0.0085 μF at zero dc volts to 0.004 μF at 22½ volts (the corresponding control current was 0.1 μA at 1½ V dc, and 3 μA at 10 V dc).

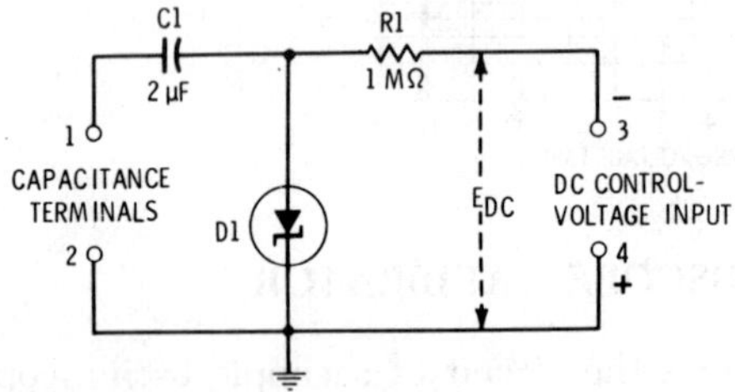

Fig. 4-24. Zener diode as voltage-variable capacitor.

4.19 ZENER DIODE AS VOLTAGE-VARIABLE RESISTOR

A reverse-biased zener diode can function as a voltage-dependent resistor (VDR) displaying a sharp transition from high resistance to low resistance. No other 2-terminal device provides so sharp a resistance change. (It will be recalled that the reverse resistance of the zener diode is very high at voltage drops below the BV value, and it is very low at voltage drops above BV.)

Fig. 4-25 shows the resistance of a Type 1N3020 zener diode ($V_Z = 10$ V) for negative voltages between 1 and 10.25 V. Note from this curve that the resistance is practically constant at 1.5 megohm between 1 V and 9.75 V, and that the resistance drops rapidly from 1.5 megohm to 100 ohms as the voltage is increased from 9.75 V to 10.25 V. At 9.75 V, the diode current is approximately 6.5 μA, and at 10.25 V it is approximately 102 mA. This represents a ratio of current change of over 15,000 to 1.

Many variations of this performance are afforded by the host of zener diodes now available, but all have the general characteristic illustrated by Fig. 4-25. For any application requiring this type of resistance/voltage curve, the zener diode is readily recommended.

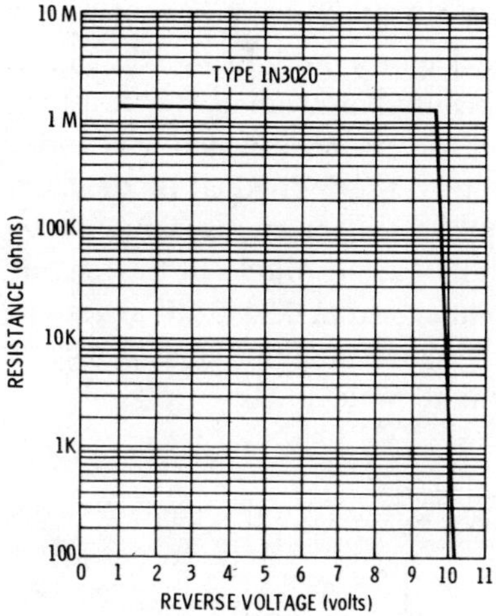

Fig. 4-25. Performance of zener diode as voltage-dependent resistor (VDR).

4.20 OSCILLOSCOPE CALIBRATOR

Fig. 4-26 shows the circuit of a simple oscilloscope voltage calibrator employing an ac connection of zener diodes. This is essentially an ac clipper (see Section 4.2). While separate, identical diodes (D1 and D2) are shown here, a single zener clipper (which contains the required two junctions) can be used in the interest of compactness and simplicity.

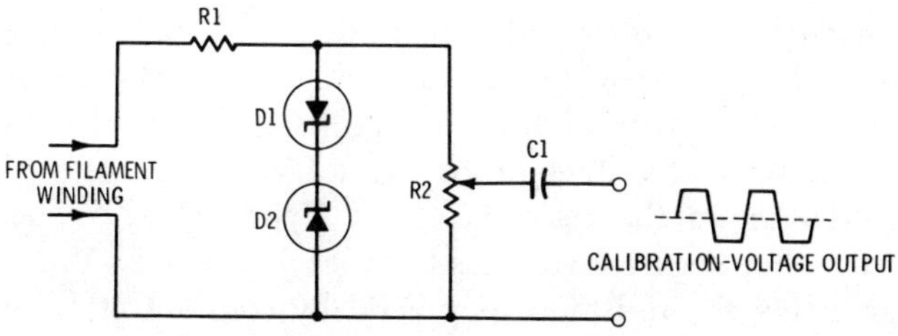

Fig. 4-26. Oscilloscope calibrator.

Ideally, the circuit is operated from one of the 6.3-V filament windings in the oscilloscope power supply, and the output voltage is a quasi-square wave. The peak-to-peak amplitude of this output voltage may be initially measured with a suitable ac voltmeter or calibrated oscilloscope, and the output potentiometer, R2, may be calibrated accordingly. Subsequently, R2 may be set for a desired peak-to-peak calibration voltage, which may be read from its dial, or

R2 may be left permanently set for one voltage (such as 1 V peak-to-peak).

The diodes and limiting resistor R1 are selected in the conventional manner for the maximum output voltage that is desired, and resistance R2 is chosen for low loading of the diode circuit. Typical constants are: D1 and D2 Type LVA 347A; R1 200 ohms, ½ watt; R2 10,000 ohms wirewound, and C1 0.1 μF.

4.21 ZENER-DIODE SUBSTITUTION BOXES

Since substitution boxes of several types (resistor, capacitor, inductor, matching transformer, etc.) are popular in practical servicing and in circuit breadboarding, the idea of the zener-diode substitution box merits some attention here. The idea itself is entirely practical, and at least one manufacturer has offered such a box; there is only the question of which diodes—out of the great number obtainable—to include for a practical working range.

With the large number of available zener diodes, the designer is hard put to make a selection for a box, in terms of which BV, V_Z, I_Z, and P ratings will be most often needed when the box is used. A further consideration is the electrical ruggedness needed in the selector switches: Many diodes have reasonably high current ratings, and the switch contacts must be able to handle these currents. The wide variety of diode ratings makes impossible a box offering the *complete* range.

Of course, an individual user, knowing his own needs, may arrange a box containing diodes having those ratings most often encountered in his work.

4.22 LOUDSPEAKER PROTECTOR

Some audiophiles have used back-to-back-connected zener diodes to protect the voice coil of an expensive loudspeaker from damage due to excess power. Varying degrees of success have been reported. The general arrangement is shown in Fig. 4-27.

The diodes must be properly chosen with respect to the correct voice-coil voltage. When the audio power delivered by the amplifier to the speaker does not exceed the maximum safe value for the speaker, the voice-coil voltage is such that the diodes operate below their BV point. The diode leg, then, offers very high resistance that has no effect on the speaker or on the fidelity of the amplifier system.

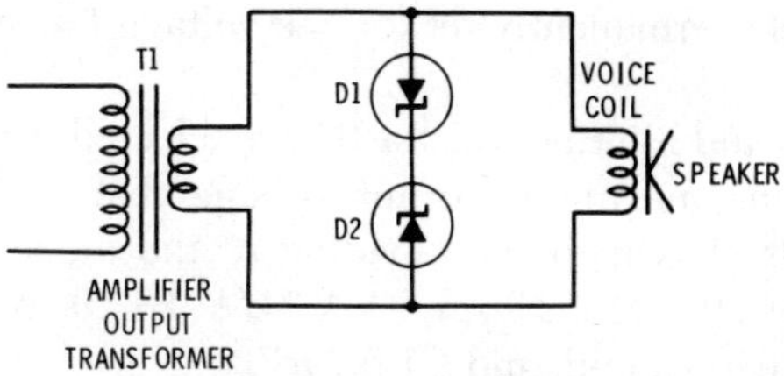

Fig. 4-27. Loudspeaker protector.

When the power exceeds the safe value, however, the diode leg breaks down, becoming a low resistance in parallel with the speaker and reducing the power applied to the speaker. This action is apt to introduce sound distortion, but this is of minor concern since the distortion serves as a signal to reduce the power at the amplifier.

This arrangement is successful only in systems in which the impedance mismatch introduced by the heavily conducting diodes is sufficient to pull down the power output of the amplifier appreciably.

4.23 ZENER DIODE AS BYPASS ELEMENT

In some circuits, a zener diode may be used in place of a bypass capacitor, and such zener bypasses are found increasingly in modern electronic circuits. The advantages of the diode over the capacitor are smallness and unlimited life. In this application, it is not the internal capacitance of the diode that is exploited (see Section 4.18), but the low ac impedance of this device, compared with its static dc resistance.

Fig. 4-28A shows a typical arrangement. Here, R1 is the resistor across which a dc voltage drop is developed by the passage of current. The current has both an ac component and a dc component,

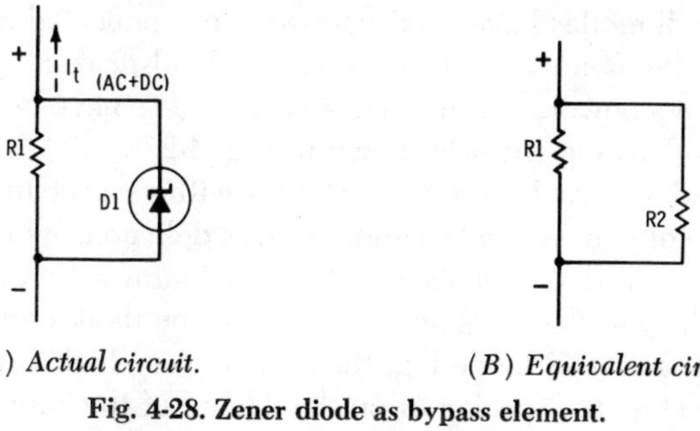

(A) *Actual circuit.* (B) *Equivalent circuit.*

Fig. 4-28. Zener diode as bypass element.

and it is desired to bypass the ac component around R1 (such a resistor is the cathode bias resistor of a vacuum tube or the emitter bias resistor of a transistor). The equivalent circuit of this arrangement is shown in Fig. 4-28B. Here, R2 represents the equivalent resistance (impedance) of the diode. The diode is operated in its zener breakdown region, so that its dc resistance, R2, has a certain value depending upon its V_Z and I_Z ratings. At this point, the ac impedance of the diode, however, is significantly lower than this dc resistance, so R2 provides a relatively easy path for alternating current around R1. As an example, the dc resistance of the diode is often several thousand ohms, while the ac impedance is of the order of 25 ohms (the reactance of a 6.3-μF capacitor at 1000 Hz).

Two main points must be kept in mind in the use of zener bypasses: (1) For dc, the total parallel resistance of the diode and resistor is effective in the circuit, so the diode must be chosen such that the total resistance equals the original desired resistance. (2) The diode is operated in its breakdown region, so the circuit in which this arrangement is installed must be able to support the zener current, I_Z. In Fig. 4-28A, the total current, I_t, equals resistor current plus zener current.

An added feature of zener bypassing is that, unlike the capacitor method, it is only slightly frequency sensitive. This should not be taken to imply that the zener diode is destined to replace the capacitor everywhere. The capacitor is a storage device, which the diode is not, and because of this, the capacitor has exclusive applications other than bypassing.

4.24 EXPANDED-SCALE VOLTMETER

A zener diode, conducting in its breakdown region, and a dc milliammeter that are connected in series provide an expanded-scale dc voltmeter. Fig. 4-29A shows a circuit of this kind giving deflections over the range 3.5 V to 10 V. Fig. 4-29B shows the calibration.

To calibrate this circuit initially, apply an accurately known 10-V dc signal to the dc input terminals, and adjust the 10,000-ohm wirewound rheostat, R1, for exact full-scale deflection of the meter. Leave the rheostat at that setting. Then, calibrate the milliammeter at other input voltage values and meter-scale points to give a chart, such as that shown in Fig. 4-29B. If desired, a special meter card may be prepared. Each circuit of this kind must be calibrated with the individual diode used. In all uses of this voltmeter, the output circuit of

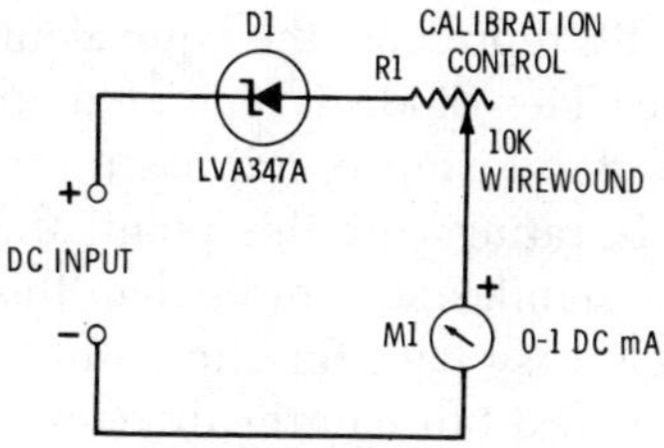

(*A*) *Circuit.*

mA	V
0.02	3.5
0.05	4.0
0.1	4.6
0.15	5.0
0.2	5.4
0.3	6.0
0.4	6.7
0.46	7.0
0.5	7.2
0.6	7.8
0.63	8.0
0.7	8.4
0.8	9.0
0.9	9.6
1.0	10

(*B*) *Calibration.*

Fig. 4-29. Expanded-scale voltmeter.

the voltage source must provide the dc return path for the diode/milliammeter circuit; that is, no blocking capacitor should be present in the path. The resistance of this return path will affect the calibration of the instrument.

The resistance of this voltmeter circuit varies from 175,000 ohms at the 3.5-V lower end of the range to 10,000 ohms at the 10-V upper end. This corresponds to a sensitivity range from 1000 ohms per volt at full scale to 50,000 ohms per volt at the low end of the scale.

4.25 COMPRESSED-SCALE VOLTMETER

In Fig. 4-30A, a Type LVA 347A zener diode (D1) and 200-ohm, ½-watt limiting resistor (R1) form a clipper circuit, which compresses the upper portion of the range of the dc voltmeter consisting of 0-1 dc milliammeter M1 and 10,000-ohm wirewound calibration-control rheostat R2. The response is shown in Fig. 4-30B; this performance is seen to be the opposite of that afforded by the expanded-

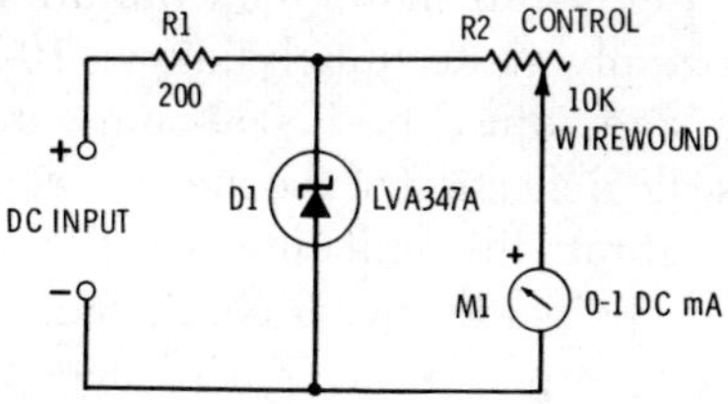

(*A*) *Circuit.*

V	mA
1	0.2
2	0.4
3	0.6
4	0.8
5	0.96
6	0.98
7	
8	
9	
10	1.0

(*B*) *Calibration.*

Fig. 4-30. Compressed-scale voltmeter.

scale voltmeter described in the preceding section. Note that the upper 4 volts of the range are compressed into the narrow space between 0.98 mA and 1 mA, and that the deflections between 6 V and 10 V are hardly readable.

To calibrate the circuit initially, apply an accurately known 10 V dc signal to the dc input terminals, and adjust the calibration-control rheostat (R2) for exact full-scale deflection of the meter. Then, calibrate the milliammeter at other input-voltage points to give a chart, such as that shown in Figure 4-30B. If desired, a special meter card may be prepared. Each circuit must be calibrated with the individual diode used. The output circuit of a voltage source tested with this voltmeter must provide the dc return path for the instrument circuit; that is, no blocking capacitor should be present in the path. The resistance of this return path will affect the calibration of the meter.

4.26 FILTER ACTION BY ZENER DIODE

Because of its ability to smooth out fluctuations in an applied dc voltage, even the most rudimentary zener-diode voltage regulator will minimize the ripple in the dc output of a rectifier. In some instances, the zener circuit is equivalent to several microfarads of filter capacitance and is a great deal smaller than the capacitor.

4.27 BASIC THRESHOLD CIRCUIT

In some circuits—such as those of automatic control devices, monitors, timers, and alarms—it is desired to have an output voltage appear only after an input voltage has exceeded a prescribed value called the *threshold*. Various means, often elaborate, are employed to obtain this action.

Fig. 4-31 shows a zener-diode voltage threshold circuit. Aside from its two-component simplicity, this arrangement provides fixed-

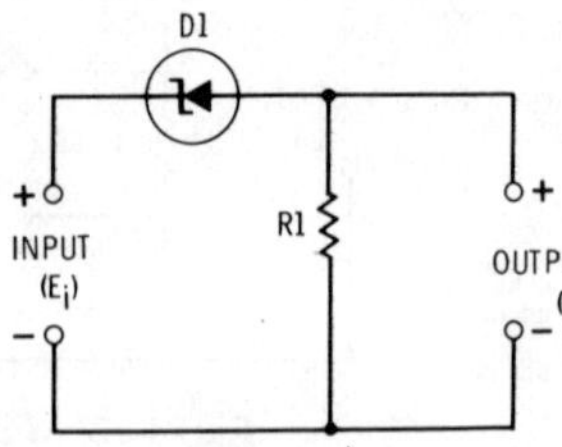

Fig. 4-31. Basic threshold circuit.

threshold operation without requiring a bias voltage, so no standby power is consumed. The single diode and resistor are easily incorporated into other circuits.

In operation, the input voltage (E_i) increases from zero or from some initial value. At all values of E_i below the BV point of zener diode D1, the diode conducts only an infinitesimal current and, for all practical purposes, may be viewed as an open circuit. Accordingly, there is virtually no voltage drop across resistor R1, and the output voltage (E_o) is effectively zero. When E_i exceeds the BV value of the diode, however, the latter breaks down and conducts heavily in its zener region. This relatively large current produces a significant voltage drop across R1, and E_o becomes evident. In this way, the BV point of the diode becomes the threshold voltage at which the circuit operates.

In order for the diode to be effective in this operation, resistance R1 must be small compared with the internal resistance (V_Z/I_Z) of the diode. And because of the voltage-divider action resulting from this relationship, output voltage E_o must always be smaller than input voltage E_i.

4.28 MULTIPLE THRESHOLD CIRCUIT

Several of the basic voltage threshold circuits described in the preceding section may be connected in parallel to provide separate outputs at different thresholds. This arrangement is shown in Fig. 4-32. Here, D1R1, D2R2, and D3R3 are the separate threshold legs, each corresponding to the simple circuit given earlier in Fig. 4-31.

The BV point of each diode determines the threshold voltage at which that leg of the circuit will deliver output voltage (E_{o1}, E_{o2}, and E_{o3}, respectively). In each instance, the output voltage continues

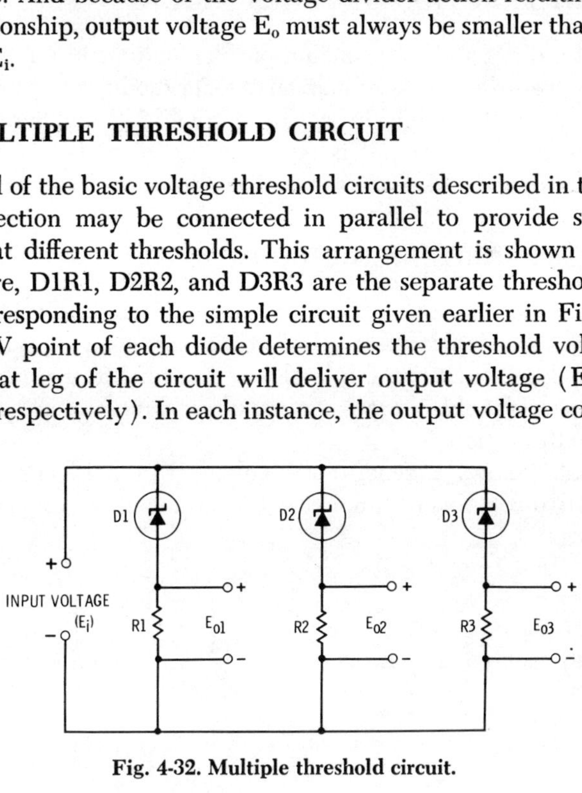

Fig. 4-32. Multiple threshold circuit.

to increase with E_i after the threshold has been exceeded; but, as explained in Section 4.27, the output voltage will always be lower than the input voltage at any point, owing to the voltage-divider action between a diode and its series resistor.

For illustration, three legs are shown in Fig. 4-32. But there is no limit to the number of such legs that can be used, so long as the input-voltage source can supply the total current for all the legs when they are operating simultaneously.

4.29 ZENER DIODES IN PROTECTED MOSFETs

In metal-oxide semiconductor field-effect transistors (MOSFETs), the thin-film insulation of the gate electrode can easily be punctured by static charges or signal overdrive, and the transistor consequently destroyed. This catastrophe is prevented in protected MOSFETs, such as Type 3N200, in which built-in zener diodes short-circuit any excess gate-signal voltage, whether due to overdrive, static, or other cause.

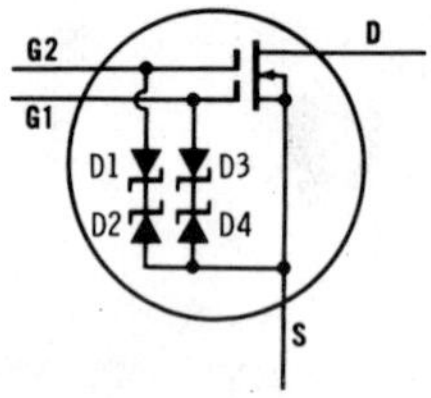

Fig. 4-33. Zener-protected MOSFET.

Fig. 4-33 shows a MOSFET of this type. In this unit, each gate is protected by a pair of back-to-back-connected zener diodes (gate 1 is protected by D3 and D4, and gate 2 by D1 and D2). These diode junctions are processed into the semiconductor subtrate along with the transistor elements. At normal gate-signal levels, these diode legs appear as extremely high resistances and so do not interfere with transistor or circuit operation. But at higher voltage, which ordinarily would damage the gate insulation, the diode legs conduct in their zener region, appear as very low resistances, and short-circuit the potentially damaging signal to the substrate electrode (usually at ground potential).

CHAPTER 5

Testing Zener Diodes

A few of the common tests for conventional diodes may also be employed with zener diodes. These include tests of ac rectification, front-to-back ratio, forward breakdown voltage, and reverse recovery time. But the characteristics that these tests appraise are usually of less interest than those of the unique reverse characteristic of the zener diode. This chapter describes a group of tests for checking zener diodes and includes only those forward characteristics which are of occasional interest in the application of these diodes.

Some transistor testers and diode testers have provision for checking zener diodes and do this job with reasonable success. In many cases, however, especially when only a few diodes must be tested at one time, the reader will set up his own test, and this chapter is offered for his guidance.

A word of warning is in order here: There is no completely reliable and foolproof way to check a zener diode with an ohmmeter, but there are several ways of damaging the diode with such a test and many ways of obtaining misleading information from the test.

5.1 ZENER VOLTAGE (V_Z)

Fig. 5-1 shows the test setup for checking zener voltage (V_Z) in terms of rated zener current (I_Z). The variable dc supply may be either a battery and potentiometer or a well-filtered, power-line-

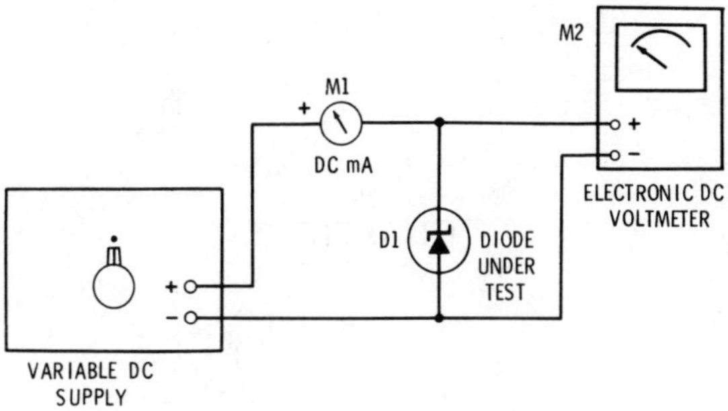

Fig. 5-1. Test setup for zener voltage (V_Z).

operated unit. The dc milliammeter (M1) should be capable of indicating the rated current (I_Z) in the upper part of its scale. The high-resistance-input electronic dc voltmeter (M2) may be either a vacuum-tube voltmeter or a transistor voltmeter. The procedure is as follows:

1. With the dc supply switched off and its output control set to zero, set up the circuit as shown in Fig. 5-1. Observe carefully all diode, supply, and meter polarities.
2. Consult the diode manufacturer's literature for the rated value of zener current I_Z.
3. Switch the dc supply on.
4. Advance the dc supply output until the deflection of milliammeter M1 corresponds to the rated I_Z from Step 2.
5. At this point, voltmeter M2 indicates the corresponding zener voltage, V_Z. This value should be checked against the diode manufacturer's literature.

5.2 ZENER CURRENT (I_Z)

Zener current (I_Z) may be checked in terms of zener voltage (V_Z) with the same test setup shown previously for zener voltage in Fig. 5-1. In this instance, as before, the variable dc supply may be either a battery and potentiometer or a well-filtered, power-line-operated unit. The dc milliammeter (M1) should have a range high enough to indicate the expected I_Z value in the upper part of its scale. The high-resistance-input electronic dc voltmeter (M2) may be either a vacuum-tube voltmeter or a transistor voltmeter. The procedure is as follows:

1. With the dc supply switched off and its output control set to zero, set up the circuit as shown in Fig. 5-1. Observe carefully all diode, supply, and meter polarities.
2. Consult the diode manufacturer's literature for the rated value of zener voltage V_Z.
3. Switch the dc supply on.
4. Advance the dc supply output until the deflection of voltmeter M2 corresponds to the rated V_Z from Step 2.
5. At this point, milliammeter M1 indicates the corresponding zener current, I_Z. This value should be checked against the diode manufacturer's literature.

5.3 REVERSE CURRENT (I_R)

Fig. 5-2 shows the test setup for checking reverse current (I_R) in terms of reverse voltage (V_R). This setup is similar to the one used for I_Z and V_Z (see Fig. 5-1), except that a sensitive dc microammeter (M1 in Fig. 5-2) is used to display I_R. The variable dc supply may be

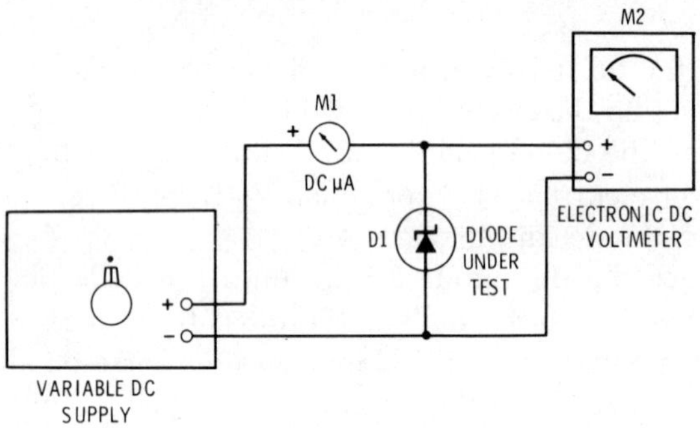

Fig. 5-2. Test setup for reverse current (I_R).

either a battery and potentiometer or a well-filtered, power-line-operated unit. The dc microammeter (M1) should be sensitive enough to indicate the expected I_R value in the upper part of its scale. The high-resistance-input electronic dc voltmeter (M2) may be either a vacuum-tube voltmeter or a transistor voltmeter. The procedure is as follows:

1. With the dc supply switched off and its output control set to zero, set up the circuit as shown in Fig. 5-2.
2. Consult the diode manufacturer's literature for the value of reverse voltage, V_R, at which I_R is specified.

3. Switch the dc supply on.
4. Advance the dc output control until the deflection of the voltmeter corresponds to the rated V_R from Step 2.
5. At this point, microammeter M1 indicates the corresponding reverse current, I_R. This value should be checked against the diode manufacturer's literature.

5.4 REVERSE VOLTAGE (V_R)

Reverse voltage (V_R) may be checked in terms of reverse current (I_R) with the same test setup shown previously for I_R in Fig. 5-2. In this instance, as before, the variable dc power supply may be either a battery and potentiometer or a well-filtered, power-line-operated unit. The dc microammeter (M1) should be sensitive enough that the rated reverse current (I_R) will be indicated in the upper part of its scale. The high-resistance-input electronic dc voltmeter (M2) may be either a vacuum-tube voltmeter or a transistor voltmeter. The procedure is as follows:

1. With the dc supply switched off and its output control set to zero, set up the circuit as shown in Fig. 5-2.
2. Consult the diode manufacturer's literature for the rated value of reverse current (I_R) for which V_R is specified.
3. Switch the dc supply on.
4. Advance the dc output control until the deflection of microammeter M1 corresponds to the rated I_R from Step 2.
5. At this point, voltmeter M2 indicates the corresponding reverse voltage, V_R. This value should be checked against the diode manufacturer's literature.

5.5 BREAKDOWN VOLTAGE (BV)

Breakdown voltage (BV) may be checked with the test setup shown in Fig. 5-1, provided that a multirange milliammeter/microammeter is used at M1 in place of the single-range milliammeter shown in the diagram. The variable dc supply may be either a battery and potentiometer or a well-filtered, power-line-operated unit. The multirange current meter (M1) should be capable of reading the low microampere current below the BV point, and also the higher current that flows through the diode after the BV point has

been passed. The high-resistance-input electronic dc voltmeter (M2) may be either a vacuum-tube voltmeter or a transistor voltmeter. The procedure is as follows:

1. With the dc supply switched off and its output control set to zero, set up the circuit as shown in Fig. 5-1. Observe carefully all diode, supply, and meter polarities.
2. Consult the diode manufacturer's literature for the rated value of breakdown voltage, BV (not all manufacturers specify this value).
3. Set the current meter, M1, to one of its microampere ranges.
4. Switch the dc supply on.
5. Slowly advance the dc output control, noting that the microamperes deflection of meter M1 increases only a small amount as the voltage is increased by a considerable amount. Be alert to switch the meter to a higher range when the current begins to increase more rapidly at the BV point.
6. Note that the current soon will begin to increase much faster as the voltage is further increased. After the BV point, the current will rise to several (or many) *milliamperes* for a very small additional increase in voltage.
7. Carefully adjust the dc output control to the point where the large current change just starts. At this point, the voltmeter (M2) indicates the breakdown voltage (BV). This value should be checked against the manufacturer's literature if it is given there.

5.6 STATIC ZENER RESISTANCE (R_Z)

The static resistance of a zener diode operating in its zener region may be calculated from the zener voltage (V_Z) and zener current (I_Z) measured according to Sections 5.1 and 5.2, respectively. After having measured V_Z and I_Z, substitute the values of these quantities in the following equation to obtain zener resistance:

$$R_Z = \frac{V_Z}{I_Z} \qquad \text{(Eq. 5-1)}$$

where,

R_Z is the zener resistance in ohms,
V_Z is the zener voltage in volts,
I_Z is the zener current in amperes.

5.7 REVERSE RESISTANCE (R_R)

The reverse resistance of a zener diode operating at a voltage far below the BV point may be calculated from the reverse current (I_R) and the reverse voltage (V_R), measured according to Sections 5.3 and 5.4, respectively. After having measured V_R and I_R, substitute these values in the following equation to obtain reverse resistance:

$$R_R = 10^6 \frac{V_R}{I_R} \qquad \text{(Eq. 5-2)}$$

where,

R_R is the zener reverse resistance in ohms,
V_R is the zener reverse voltage in volts,
I_R is the zener reverse (leakage) current in microamperes.

5.8 DYNAMIC IMPEDANCE (Z_Z)

Fig. 5-3 shows a test setup for checking the dynamic zener impedance (Z_Z) of a zener diode at 60 Hz. In this arrangement, a 60-Hz current, supplied from the power line through a step-down (filament-type) transformer (T1) and adjusted by means of a Variac, is superimposed upon the dc zener current. The latter is supplied by the

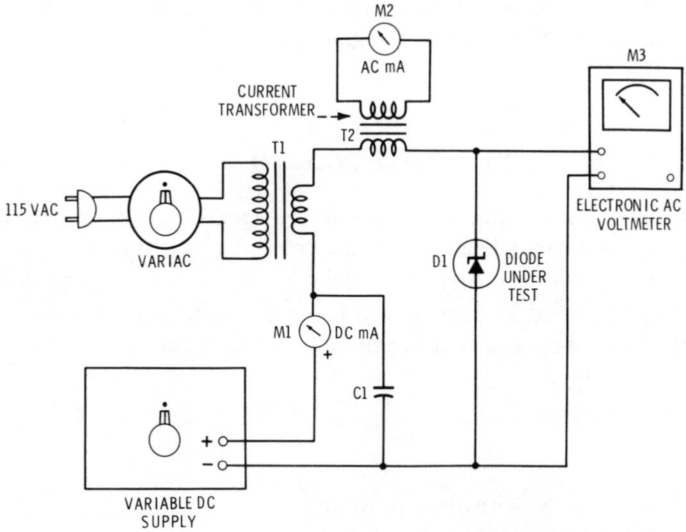

Fig. 5-3. Test setup for dynamic impedance (Z_Z).

adjustable dc supply, which may be a battery and potentiometer or a well-filtered, power-line-operated unit.

The direct current is indicated by a dc milliammeter (M1); the alternating current is indicated by an ac milliammeter (M2) isolated from the dc component by means of a suitable current transformer, T2. Capacitor C1 (of the order of 10 μF) bypasses the alternating current around the dc supply and dc milliammeter. The ac voltage drop across diode D1 under test is indicated by the electronic ac voltmeter (M3), which may be either a vacuum-tube voltmeter or a transistor voltmeter. The procedure is as follows:

1. With the dc supply switched off (and its output control set to zero) and with the Variac set to zero (and unplugged from the power line), set up the circuit as in Fig. 5-3.
2. Switch the dc supply on.
3. Consult the diode manufacturer's literature for the rated value of zener current (I_Z) for the diode under test.
4. Advance the dc output until dc milliammeter M1 indicates the value of zener current, I_Z, obtained in Step 3.
5. Plug the Variac into the power line.
6. Advance the Variac until the peak value of the deflection of ac milliammeter M2 equals 1/10 of the zener current (I_Z) obtained in Step 4. Record the rms value of this current as I_{ac}.
7. At this point, observe the deflection of voltmeter M3, and record this rms value as V_{ac}.
8. In the following equation, substitute the values obtained for I_{ac} and V_{ac} to obtain dynamic impedance:

$$Z_Z = \frac{V_{ac}}{I_{ac}} \qquad \text{(Eq. 5-3)}$$

where,

Z_Z is the zener impedance in ohms,
V_{ac} is the ac voltage in rms volts,
I_{ac} is the ac current in rms amperes.

5.9 NOISE DENSITY (N_D)

Fig. 5-4 shows a test setup employed in the industry to check zener-diode noise density (N_D). In this arrangement, the variable dc supply may be either a battery and potentiometer or an extremely well-filtered and low-noise-output, power-line-operated unit. Resistance R1 is chosen high enough to make the dc supply essentially

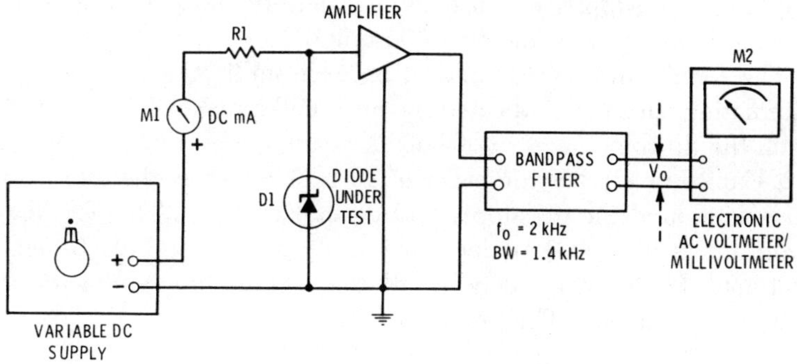

Fig. 5-4. Test setup for noise density (N_D).

a constant-current source. The amplifier must have an accurately known ac voltage gain (A) and an extremely low noise output. The bandpass filter has a center frequency (f_o) of 2 kHz and a bandwidth of 1.4 kHz. The dc milliammeter (M1) must be capable of indicating the zener current, I_Z, of the diode (D1) under test in the upper part of its scale. The electronic ac voltmeter/millivoltmeter may be either a vacuum-tube voltmeter or transistor voltmeter. The procedure is as follows:

1. With the dc supply switched off and its output control set to zero, set up the circuit as in Fig. 5-4.
2. Switch the dc supply on.
3. Consult the diode manufacturer's literature for the rated value of zener current (I_Z) for the diode under test.
4. Adjust the dc output control until milliammeter M1 indicates the value of zener current found in Step 3.
5. At this point, the voltmeter (M2) indicates the noise voltage V_o.
6. Calculate the noise density by means of the following equation:

$$N_D = \frac{V_o}{37.42A} \qquad \text{(Eq. 5-4)}$$

where,

N_D is the noise density in volts per square root hertz,
V_o is the noise voltage in rms volts,
A is the voltage amplification of the system.

Equation 5-4 assumes that the bandwidth of the filter is exactly 1.4 kHz. If the bandwidth (BW) is some other figure, the equation becomes $N_D = V_o/(A\sqrt{BW})$, where BW is in Hz.

5.10 FORWARD CURRENT (I_F)

Fig. 5-5 shows the test setup for checking forward current (I_F) in terms of rated forward voltage (V_F). In this arrangement, the variable dc supply may be either a battery and potentiometer or a well-filtered, power-line-operated unit. The dc milliammeter (M1) should be capable of indicating the expected current (I_F) in the upper

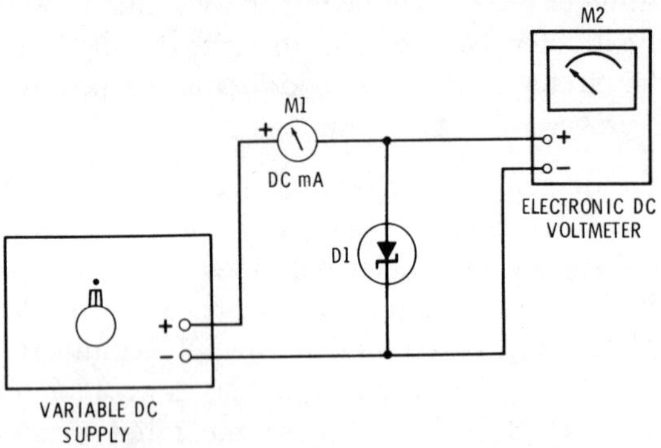

Fig. 5-5. Test setup for forward current (I_F).

part of its scale. The high-resistance-input electronic dc voltmeter (M2) may be either a vacuum-tube voltmeter or transistor voltmeter. Note that in this test the diode (D1) is connected so that its anode is biased positive by the dc supply. The procedure is as follows:

1. With the dc supply switched off and its output control set to zero, set up the circuit as shown in Fig. 5-5. Observe carefully all diode, supply, and meter polarities.
2. Consult the diode manufacturer's literature for the rated value of forward voltage (V_F). Not all manufacturers specify this voltage; when the V_F value is not available, 1 V will probably be satisfactory.
3. Switch the dc supply on.
4. Advance the dc output until the deflection of voltmeter M2 corresponds to the rated V_F from Step 2.
5. At this point, milliammeter M1 indicates the corresponding forward current, I_F. This value should be checked against the diode manufacturer's figure, when the latter is available.

5.11 FORWARD VOLTAGE (V_F)

Forward voltage (V_F) may be checked in terms of forward current (I_F) with the same test setup shown previously for forward current in Fig. 5-5. In this instance, as before, the variable dc supply may be either a battery and potentiometer or a well-filtered, power-line-operated unit. The dc milliammeter (M1) should have a range high enough to indicate the rated forward current (I_F) of the diode under test in the upper part of its scale. The high-resistance-input electronic dc voltmeter (M2) may be either a vacuum-tube voltmeter or a transistor voltmeter. Note that in this test the diode (D1) is connected in the circuit so that its anode is biased positive by the dc supply. The procedure is as follows:

1. With the dc supply switched off and its output control set to zero, set up the circuit as shown in Fig. 5-5. Observe carefully all diode, supply, and meter polarities.
2. Consult the diode manufacturer's literature for the rated test value of forward current (I_F). Not all manufacturers specify this current; when it is not available, the reader must select a test value which will not cause the rated maximum power dissipation of the diode to be exceeded.
3. Switch the dc supply on.
4. Advance the dc output until the deflection of milliammeter M1 corresponds to the rated I_F from Step 2.
5. At this point, voltmeter M2 indicates the corresponding forward voltage, V_F. This value should be compared with the diode manufacturer's figure, when the latter is available.

5.12 STATIC FORWARD RESISTANCE (R_F)

The static resistance of a zener diode operating in its forward region may be calculated from the forward current (I_F) and the forward voltage (V_F), measured according to Sections 5.10 and 5.11, respectively. After having measured I_F and V_F, substitute the values of these quantities in the following equation to obtain the forward resistance:

$$R_F = \frac{V_F}{I_F} \qquad \text{(Eq. 5-5)}$$

where,

R_F is the forward resistance in ohms,
V_F is the forward voltage in volts,
I_F is the forward current in amperes.

Since the forward resistance is nonlinear, its value may be interesting to determine at several different values of current or voltage.

5.13 FORWARD POWER DISSIPATION (P_F)

The forward power dissipation of a zener diode may be calculated from the forward current (I_F) and the forward voltage (V_F), measured according to Sections 5.10 and 5.11, respectively. After having measured I_F and V_F, substitute the values of these quantities in the following equation to obtain forward power dissipation:

$$P_F = V_F I_F \qquad \text{(Eq. 5-6)}$$

where,
P_F is the forward power in watts,
V_F is the forward voltage in volts,
I_F is the forward current in amperes.

5.14 ZENER POWER DISSIPATION (P_Z)

The power dissipation of a zener diode operating in its reverse-breakdown (zener) region may be calculated from the zener voltage (V_Z) and the zener current (I_Z), measured according to Sections 5.1 and 5.2, respectively. After having measured V_Z and I_Z, substitute the values of these quantities in the following equation to obtain zener power:

$$P_Z = V_Z I_Z \qquad \text{(Eq. 5-7)}$$

where,
P_Z is the zener power in watts,
V_Z is the zener voltage in volts,
I_Z is the zener current in amperes.

5.15 VISUAL TEST OF DIODE

Figs. 5-6A and 5-7A show test setups for oscilloscope display of a zener-diode operating curve. With the first circuit, the display (Fig. 5-6B) shows the entire characteristic—that is, both the forward and reverse portions of the diode conduction curve. With the second circuit, however, the display (Fig. 5-7B) is only the reverse portion of the conduction curve. In this latter case, the unneeded forward characteristic is suppressed by silicon rectifier D1 (Fig. 5-7A), which supplies only the negative half-cycle of test voltage to the zener diode (D2) under test. In each instance, the oscilloscope is a voltage-

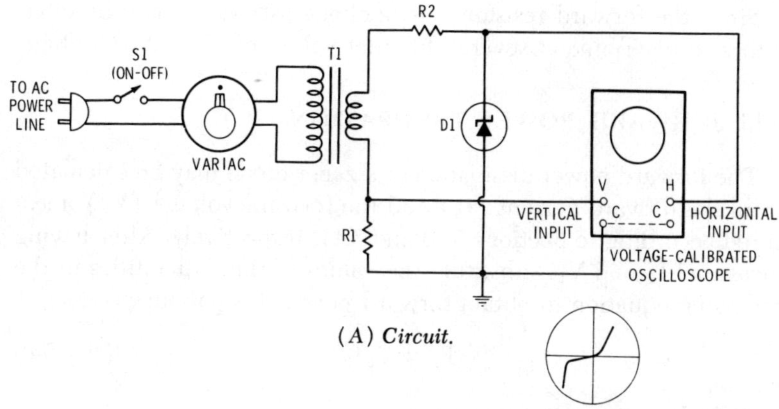

(A) *Circuit.*

(B) *Display.*

Fig. 5-6. Visual test setup (complete display).

calibrated instrument, so the current and voltage values may be read along the vertical and horizontal axes of the screen, respectively.

In each circuit, the following conditions hold: (1) Transformer T1 must be selected for a peak secondary voltage corresponding to the highest reverse voltage that is to be applied to the diode under test, and this voltage is closely set by means of the Variac. (2) Re-

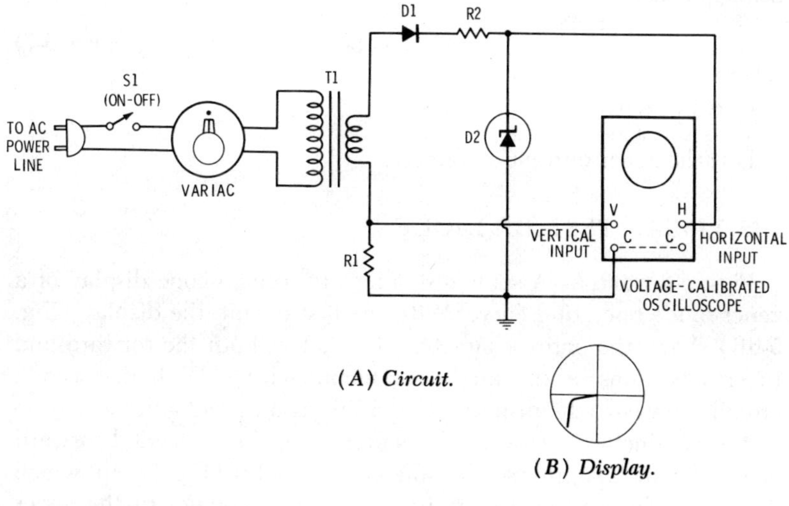

(A) *Circuit.*

(B) *Display.*

Fig. 5-7. Visual test setup (reverse display).

sistor R2 is a current-limiting resistor of the type usually employed in conjunction with a zener diode. This resistor has the additional function, however, that in the circuit in Fig. 5-6 it must also limit the forward current passed by the diode during the positive half-cycle of test voltage. (3) The voltage across the diode is applied to the horizontal channel of the oscilloscope; thus, the horizontal deflection is proportional to V_Z in the Fig. 5-7 circuit and to V_Z and V_F in the Fig. 5-6 circuit. (4) The current through the diode flows through low resistance R1 (typically 1 ohm), and the voltage drop across this resistor (directly proportional to the diode current) is applied to the vertical channel of the oscilloscope. Thus, the vertical deflection is proportional to I_Z in the Fig. 5-7 circuit and to I_Z and I_F in the Fig. 5-6 circuit. Since resistance R1 is very small, it introduces only negligible voltage loss in the circuit. When R1 = 1 ohm, the diode current may be read directly on the vertical scale of the voltage-calibrated oscilloscope, since $I = V/R$, and in this case $I = V/1$.

5.16 TEMPERATURE COEFFICIENT (TC OR θ)

Any one of the zener-diode characteristics (BV, V_Z, I_Z, Z_Z, I_R, and so on) may be measured over a desired temperature range, and the temperature coefficient of that characteristic (e.g., temperature coefficient of breakdown voltage BV) determined from the change exhibited by the characteristic during the experienced temperature change. For this purpose, the diode must be operated inside a reliable, adjustable temperature chamber, with short leads extending to the external measuring instruments. The temperature of the chamber must be held at each point long enough for the diode temperature to stabilize at that point.

The temperature coefficient is calculated from the ratio of the change in the characteristic to the change in temperature, and it may be expressed in units per degree (e.g., volts per degree), in percent of the initial value of the characteristic, or in parts per million per degree.

APPENDIX

Abbreviations Used in This Book

A—Ampere(s); amplification (gain)
ac—Alternating-current
af—Audio-frequency
AMP—Amplifier
B—Battery
BV—Breakdown voltage
BW—Bandwidth
C—Capacitance, capacitor; Celsius (temperature)
C_J—Junction capacitance
D—Diode; drain (electrode of field-effect transistor)
dc—Direct-current
E—Voltage, volts
E_c—Bias voltage
e.g.—For example
E_i—Input voltage
E_o—Output voltage
FETvm—Field-effect-transistor voltmeter
FL—Full load
f_o—Center frequency
G—Gate (electrode of field-effect transistor)

Hz—Hertz

I—Current

I_{ac}—Ac current

i.e.—That is

I_F—Forward current

i-f—Intermediate-frequency

I_L—Load current

I_R—Reverse (leakage) current

I_t—Total current

I_Z—Zener current

I_{ZT}—Zener current (test value)

K—Times 1000

kHz—Kilohertz

M—Meter

mA—Milliampere(s)

meg—Megohm(s)

MOSFET—Metal-oxide semiconductor field-effect transistor

N_D—Noise density

NL—No load

npn—Transistor structure with n-type emitter and collector and p-type base

P—Power

P_D—Power dissipation

P_{DM}—Maximum allowable power dissipation

P_F—Forward power

pF—Picofarad(s)

pn—Positive-negative (referring to a semiconductor junction between p- and n-doped material)

pnp—Transistor structure with p-type emitter and collector and n-type base

P_Z—Zener power

Q—Transistor symbol

R—Resistance, resistor

RECT—Rectifier

R_F—Forward resistance

rfc—Radio-frequency choke

R_m—Meter resistance
rms—Root mean square
R_R—Reverse resistance
R_t—Total resistance
RY—Relay
R_Z—Zener resistance
S—Source (electrode of field-effect transistor)
T—Transformer; temperature
TC—Temperature coefficient
V—Vacuum tube; volt(s), voltage
V_{ac}—Ac voltage
VDR—Voltage-dependent resistor
V_F—Forward voltage
V_o—Noise voltage
V_R—Reverse voltage
VR—Voltage regulator, voltage regulation
vtvm—Vacuum-tube voltmeter
V_Z—Zener voltage
W—Watt(s)
ww—Wirewound
X_c—Capacitive reactance
Z_c—Capacitor impedance
Z_Z—Zener impedance

GREEK LETTERS

Δ—Increment
μA—Microampere(s)
μF—Microfarads(s)
μV—Microvolt(s)
Ω—Ohm(s)
θ—Temperature coefficient

Index

Z